INVESTIGATIONS GUIDE

Animals Two by Two

FOSS Next Generation

Full Option Science System
Developed at the Lawrence Hall of Science, University of California, Berkeley
Published and Distributed by Delta Education

FOSS Lawrence Hall of Science Team
Larry Malone and Linda De Lucchi, FOSS Project Codirectors and Lead Developers
Kathy Long, FOSS Assessment Director; David Lippman, Program Manager; Carol Sevilla, Publications Design Coordinator; Susan Stanley, Illustrator; John Quick, Photographer
FOSS Curriculum Developers: Brian Campbell, Teri Lawson, Alan Gould, Susan Kaschner Jagoda, Ann Moriarty, Jessica Penchos, Kimi Hosoume, Virginia Reid, Joanna Snyder, Erica Beck Spencer, Joanna Totino, Diana Velez, Natalie Yakushiji
Susan Ketchner, Technology Project Manager
FOSS Technology Team: Dan Bluestein, Christopher Cianciarulo, Matthew Jacoby, Kate Jordan, Frank Kusiak, Nicole Medina, Jonathan Segal, Dave Stapley, Shan Tsai

Delta Education Team
Bonnie A. Piotrowski, Editorial Director, Elementary Science
Project Team: Jennifer Apt, Sandra Burke, Joann Hoy, Kristen Mahoney, Jennifer McKenna, Angela Miccinello

Thank you to all FOSS Grades K-5 Trial Teachers
Heather Ballard, Wilson Elementary, Coppell, TX; Mirith Ballestas De Barroso, Treasure Forest Elementary, Houston, TX; Terra L. Barton, Harry McKillop Elementary, Melissa, TX; Rhonda Bernard, Frances E. Norton Elementary, Allen, TX; Theresa Bissonnette, East Millbrook Magnet Middle School, Raleigh, NC; Peter Blackstone, Hall Elementary School, Portland, ME; Tiffani Brisco, Seven Hills Elementary, Newark, TX; Darrow Brown, Lake Myra Elementary School, Wendell, NC; Heather Callaghan, Olive Chapel Elementary, Apex, NC; Katie Cannon, Las Colinas Elementary, Irving, TX; Elaine M. Cansler, Brassfield Road Elementary School, Raleigh, NC; Kristy Cash, Wilson Elementary, Coppell, TX; Monica Coles, Swift Creek Elementary School, Raleigh, NC; Shirley Conner, Ocean Avenue Elementary School, Portland, ME; Sally Connolly, Cape Elizabeth Middle School, Cape Elizabeth, ME; Melissa Cook-Airhart, Harry McKillop Elementary, Melissa, TX; Melissa Costa, Olive Chapel Elementary, Apex, NC; Hillary P. Croissant, Harry McKillop Elementary, Melissa, TX; Rene Custeau, Hall Elementary School, Portland, ME; Nancy Davis, Martha and Josh Morriss Mathematics and Engineering Elementary School, Texarkana, TX; Nancy Deveneau, Wilson Elementary, Coppell, TX; Karen Diaz, Las Colinas Elementary, Irving, TX; Marlana Dumas, Las Colinas Elementary, Irving, TX; Mary Evans, R.E. Good Elementary School, Carrollton, TX; Jacquelyn Farley, Moss Haven Elementary, Dallas, TX; Corinna Ferrier, Oak Forest Elementary, Humble, TX; Allison Fike, Wilson Elementary, Coppell, TX; Barbara Fugitt, Martha and Josh Morriss Mathematics and Engineering Elementary School, Texarkana, TX; Colleen Garvey, Farmington Woods Elementary, Cary, NC; Judy Geller, Bentley Elementary School, Oakland, CA; Erin Gibson, Las Colinas Elementary, Irving, TX; Kelli Gobel, Melissa Ridge Intermediate School, Melissa, TX; Dollie Green, Melissa Ridge Intermediate School, Melissa, TX; Brenda Lee Harrigan, Bentley Elementary School, Oakland, CA; Cori Harris, Samuel Beck Elementary, Trophy Club, TX; Kim Hayes, Martha and Josh Morriss Mathematics and Engineering Elementary School, Texarkana, TX; Staci Lynn Hester, Lacy Elementary School, Raleigh, NC; Amanda Hill, Las Colinas Elementary, Irving, TX; Margaret Hillman, Ocean Avenue Elementary School, Portland, ME; Cindy Holder, Oak Forest Elementary, Humble, TX; Sarah Huber, Hodge Road Elementary, Knightdale, NC; Susan Jacobs, Granger Elementary, Keller, TX; Carol Kellum, Wallace Elementary, Dallas, TX; Jennifer A. Kelly, Hall Elementary School, Portland, ME; Brittani Kern, Fox Road Elementary, Raleigh, NC; Jodi Lay, Lufkin Road Middle School, Apex, NC; Melissa Lourenco, Lake Myra Elementary School, Wendell, NC; Ana Martinez, RISD Academy, Dallas, TX; Shaheen Mavani, Las Colinas Elementary, Irving, TX; Mary Linley McClendon, Math Science Technology Magnet School, Richardson, TX; Adam McKay, Davis Drive Elementary, Cary, NC; Leslie Meadows, Lake Myra Elementary School, Wendell, NC; Anne Mechler, J. Erik Jonsson Community School, Dallas, TX; Anne Miller, J. Erik Jonsson Community School, Dallas, TX; Shirley Diann Miller, The Rice School, Houston, TX; Keri Minier, Las Colinas Elementary, Irving, TX; Stephanie Renee Nance, T.H. Rogers Elementary, Houston, TX; Cynthia Nilsen, Peaks Island School, Peaks Island, ME; Elizabeth Noble, Las Colinas Elementary, Irving, TX; Courtney Noonan, Shadow Oaks Elementary School, Houston, TX; Sarah Peden, Aversboro Elementary School, Garner, NC; Carrie Prince, School at St. George Place, Houston, TX; Marlaina Pritchard, Melissa Ridge Intermediate School, Melissa, TX; Alice Pujol, J. Erik Jonsson Community School, Dallas, TX; Claire Ramsbotham, Cape Elizabeth Middle School, Cape Elizabeth, ME; Paul Rendon, Bentley Elementary, Oakland, CA; Janette Ridley, W.H. Wilson Elementary School, Coppell, TX; Kristina (Crickett) Roberts, W.H. Wilson Elementary School, Coppell, TX; Heather Rogers, Wendell Creative Arts & Science Magnet Elementary School, Wendell, NC; Alissa Royal, Melissa Ridge Intermediate School, Melissa, TX; Megan Runion, Olive Chapel Elementary, Apex, NC; Christy Scheef, J. Erik Jonsson Community School, Dallas, TX; Samrawit Shawl, T.H. Rogers School, Houston, TX; Nicole Spivey, Lake Myra Elementary School, Wendell, NC; Ashley Stephenson, J. Erik Jonsson Community School, Dallas, TX; Jolanta Stern, Browning Elementary School, Houston, TX; Gale Stimson, Bentley Elementary, Oakland, CA; Ted Stoeckley, Hall Middle School, Larkspur, CA; Cathryn Sutton, Wilson Elementary, Coppell, TX; Camille Swander, Ocean Avenue Elementary School, Portland, ME; Brandi Swann, Westlawn Elementary School, Texarkana, TX; Robin Taylor, Arapaho Classical Magnet, Richardson, TX; Michael C. Thomas, Forest Lane Academy, Dallas, TX; Jomarga Thompkins, Lockhart Elementary, Houston, TX; Mary Timar, Madera Elementary, Lake Forest, CA; Helena Tongkeamha, White Rock Elementary, Dallas, TX; Linda Trampe, J. Erik Jonsson Community School, Dallas, TX; Charity VanHorn, Fred A. Olds Elementary, Raleigh, NC; Kathleen VanKeuren, Lufkin Road Middle School, Apex, NC; Valerie Vassar, Hall Elementary School, Portland, ME; Megan Veron, Westwood Elementary School, Houston, TX; Mary Margaret Waters, Frances E. Norton Elementary, Allen, TX; Stephanie Robledo Watson, Ridgecrest Elementary School, Houston, TX; Lisa Webb, Madisonville Intermediate, Madisonville, TX; Matt Whaley, Cape Elizabeth Middle School, Cape Elizabeth, ME; Nancy White, Canyon Creek Elementary, Austin, TX; Barbara Yurick, Oak Forest Elementary, Humble, TX; Linda Zittel, Mira Vista Elementary, Richmond, CA

Photo Credits: © Vlad61/Shutterstock; © photka/Shutterstock (cover); © gary718/Shutterstock; © Laurie Meyer; © John Quick; © Erica Beck Spencer

Published and Distributed by Delta Education, a member of the School Specialty Family
The FOSS program was developed in part with the support of the National Science Foundation grant nos. MDR-8751727 and MDR-9150097. However, any opinions, findings, conclusions, statements, and recommendations expressed herein are those of the authors and do not necessarily reflect the views of NSF. FOSSmap was developed in collaboration between the BEAR Center at UC Berkeley and FOSS at the Lawrence Hall of Science.

Copyright © 2018 by The Regents of the University of California

Standards cited herein from NGSS Lead States. 2013. *Next Generation Science Standards: For States, By States*. Washington, DC: The National Academies Press. Next Generation Science Standards is a registered trademark of Achieve. Neither Achieve nor the lead states and partners that developed the Next Generation Science Standards was involved in the production of, and does not endorse, this product.

All rights reserved. Any part of this work may not be reproduced or transmitted in any form or by any means, electronic or mechanical, including photocopying and recording, or by an information storage or retrieval system without prior written permission. For permission please write to: FOSS Project, Lawrence Hall of Science, University of California, Berkeley, CA 94720 or foss@berkeley.edu.

Animals Two by Two
Investigations Guide, 1487583
978-1-62571-417-6
Printing 4 – 6/2017
Webcrafters, Madison, WI

INVESTIGATIONS GUIDE

Animals Two by Two

TABLE OF CONTENTS

Overview 1
Framework and NGSS 27
Materials 43
Technology 55

Investigation 1: Goldfish and Guppies 65
Part 1: The Structure of Goldfish 78
Part 2: Caring for Goldfish 86
Part 3: Goldfish Behavior 91
Part 4: Comparing Guppies to Goldfish 96
Part 5: Comparing Schoolyard Birds 104

Investigation 2: Water and Land Snails 119
Part 1: Observing Water Snails 130
Part 2: Shells 137
Part 3: Land Snails 142

Investigation 3: Big and Little Worms 157
Part 1: The Structure of Redworms 168
Part 2: Redworm Behavior 174
Part 3: Comparing Redworms to Night Crawlers 182

Investigation 4: Pill Bugs and Sow Bugs 193
Part 1: Isopod Observations 202
Part 2: Identifying Isopods 208
Part 3: Isopod Movement 215
Part 4: Animals Living Together 224

Assessment 233

Welcome to FOSS® Next Generation™

Getting Started with FOSS Next Generation for Grades K–2

Whether you're new to hands-on science or a FOSS veteran, you'll be up and running in no time and ready to lead your students on a fantastic voyage through the wonders of the natural and designed world.

Watch our short video series or browse the next few pages to get started!

Getting Started with FOSS: Meet Your Module video

Scan here or visit deltaeducation.com/goFOSS

Three-Dimensional Active Science

It's time to experience the three dimensions of the NGSS—**disciplinary core ideas**, **crosscutting concepts**, and **science and engineering practices**. Engage in rich investigations that immerse your students in real-world applications of important scientific phenomenon, supported by just-in-time teaching tips and strategies.

Getting Started with Your Equipment Kit

Meet Your FOSS Module!

Your FOSS module includes one or more large boxes, called drawers, and two smaller boxes for the Teacher Toolkit, student books, and other equipment. Each drawer has a label on the front listing its contents. Your packing list is always in Drawer 1.

Permanent Equipment

Your equipment kit includes enough permanent equipment for up to 8 groups (32 students). This equipment is classroom-tested and expected to last 7–10 years.

Consumable Equipment

Your kit also includes consumable materials for three class uses. Convenient refill kits provide materials for three additional uses and are available through Delta Education.

Easy Set-up and Clean-up!

FOSS Next Generation equipment drawers are packed by investigation to facilitate prep and to make packing up for the next use a snap!

Drawer sections include:

- Unique materials needed for one investigation
- Common equipment used in multiple investigations
- Consumable materials—when it's empty you know it's time to refill!

Order Refills Online

deltaeducation.com/refillcenter

Live Organisms

Some investigations require live organisms. Schools are encouraged to purchase these organisms from a local biological supply company to minimize both transit time and the impact of adverse weather on the health of the organisms.

If living material cards are purchased from Delta Education, they will be shipped separately in a green and white envelope. Keep these cards in a safe place until it's time to redeem them for the investigation.

Call Delta Education at 800-258-1302 at least three weeks before you need your organisms.

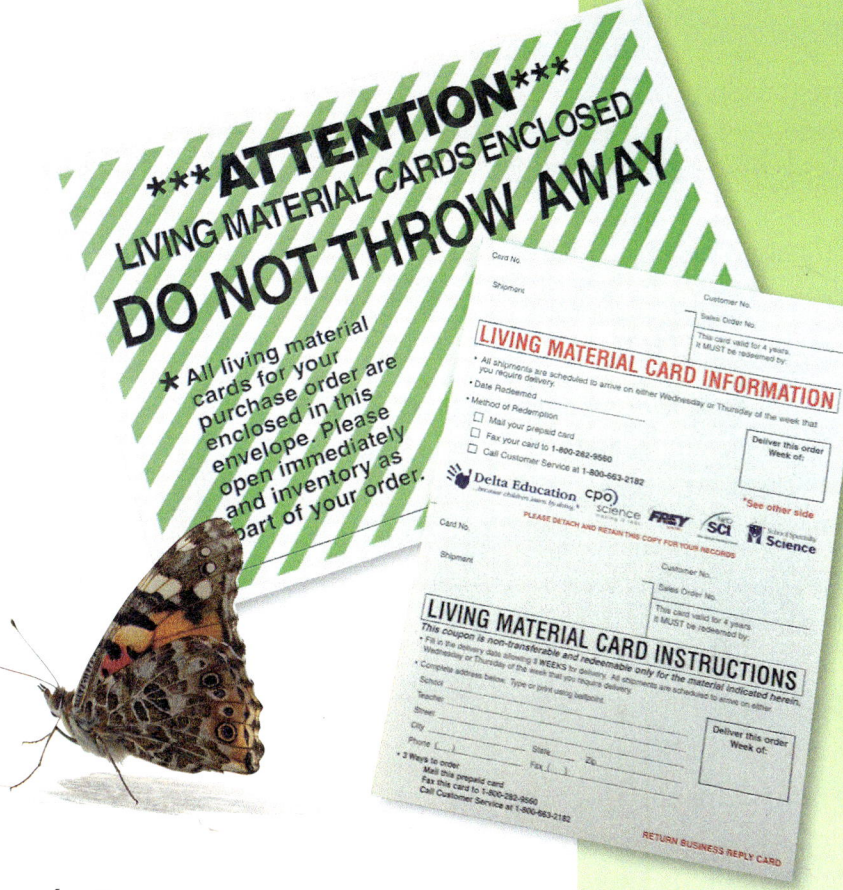

Premium Student eBook Access

If your school purchased a premium class license for the *FOSS Science Resources* student eBook, your access codes will be shipped separately in a blue and white striped envelope. Use this access code on FOSSweb to unlock student eBook access.

Getting Started with Your Teacher Toolkit

The Teacher Toolkit is the most important part of the FOSS program. There are three parts of the Teacher Toolkit—the **Investigations Guide**, **Teacher Resources**, and the student **Science Resources** Book. It's here that all the wisdom and experience from years of research and classroom development comes together to support teachers with lesson facilitation and in-depth strategies for taking investigations to the next level.

1. Investigations Guide

The **Investigations Guide** is your roadmap to prepare for and lead the FOSS investigations. Chapters are tabbed for easy access to important module information.

The module **Overview** gives you a high-level look at the 10–12 weeks of instruction in each module including a summary matrix, schedule for the module, and product support contacts.

Framework and the NGSS provides a complete overview of NGSS connections, learning progressions, and background to support the conceptual framework for the module.

The **Materials** chapter is a must-read resource that helps you get your student equipment ready for first-time use and shares helpful tips for getting your classroom ready for FOSS.

The **Technology** chapter provides an overview for each digital resource in the module and gets you up and running on FOSSweb.com, complete with technical support.

Each **Investigation** includes an At-a-Glance overview, science background content with NGSS connections, and in-depth guidance for preparing and facilitating instruction.

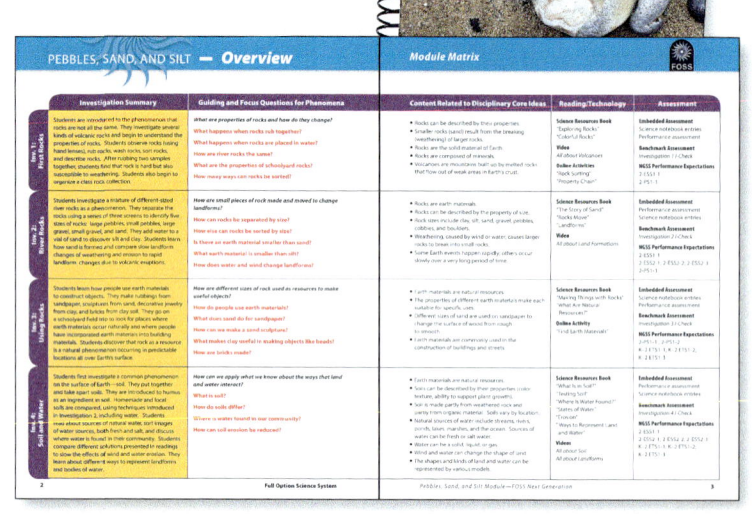

Module matrix

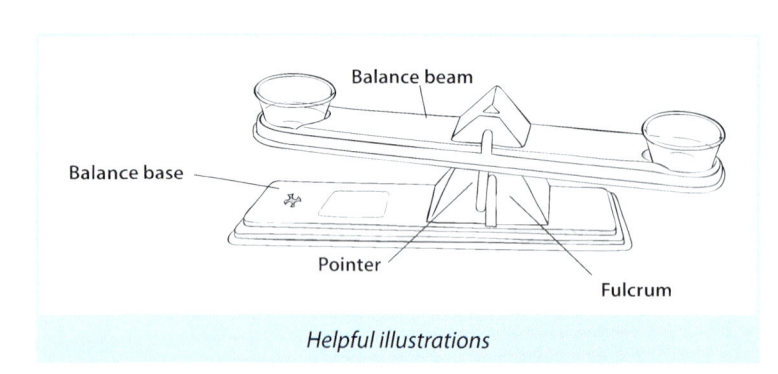

Helpful illustrations

The At-a-Glance chart includes:

- Summaries and pacing for investigation scheduling
- Focus questions for investigative phenomena
- Connections to disciplinary core idea
- Reading, writing, and technology integration opportunities
- Embedded and benchmark assessments

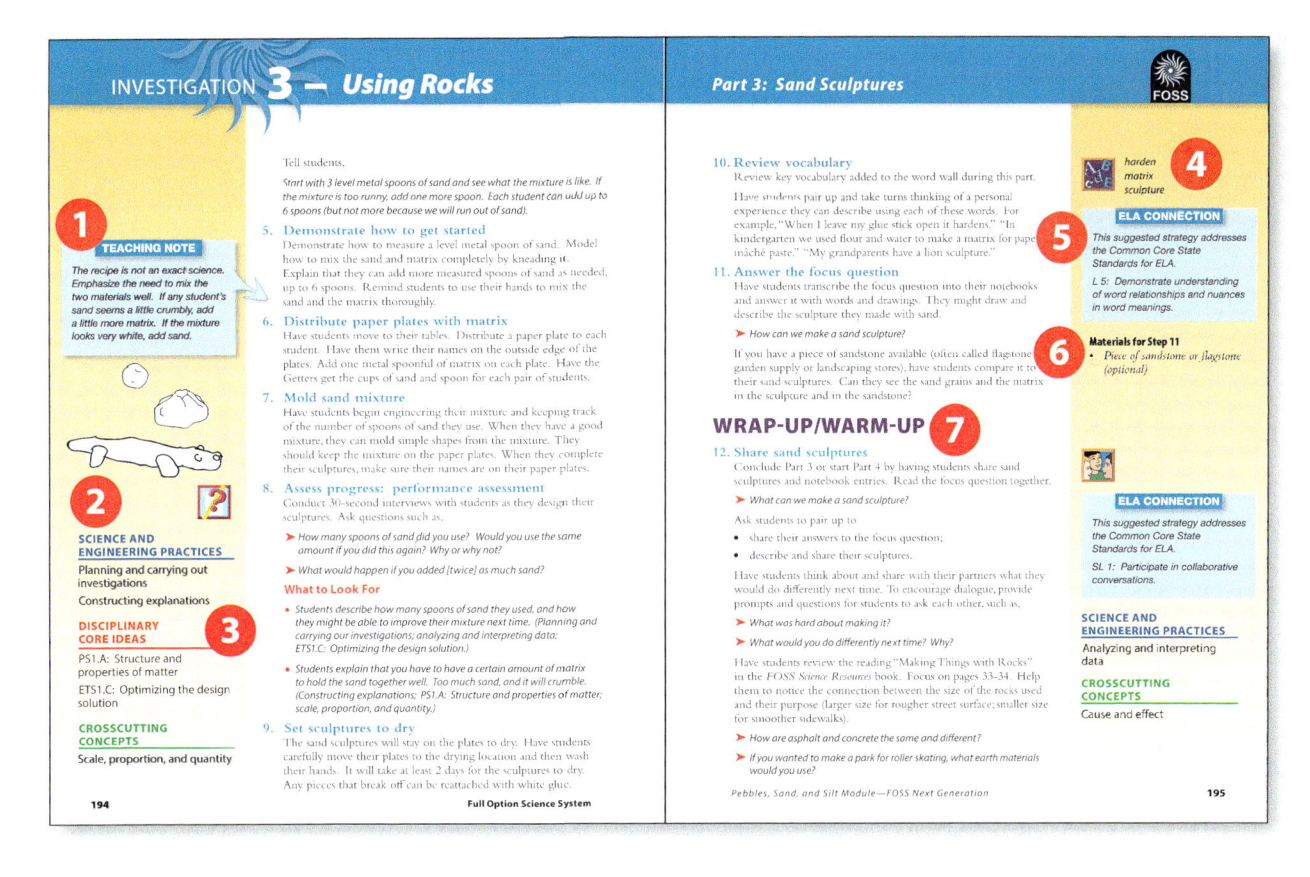

FOSS investigations provide the right support, when you need it with point-of-use guidance.

1. Teaching notes from real classrooms
2. Key three-dimensional highlights
3. Embedded assessment "What to Look For" in grades 1–2
4. Vocabulary review
5. Strategies to support English Language Arts
6. Materials used in the current steps
7. Guiding questions to help students make connections

The **Assessment** chapter gives you an in-depth look at the research-based components of the FOSS Assessment System, guidance on assessing for the NGSS, and generalized next-step strategies to use in your classroom. Find duplication masters, assessment charts, coding guides, and specific next-step strategies on FOSSweb.com.

Getting Started with Your Teacher Toolkit

2. Teacher Resources

Your *Investigations Guide* tells you how to facilitate each investigation of a module. The **Teacher Resources** provides guidance on how to do it at your grade level across three modules throughout the year with effective practices and strategies derived from extensive field-testing.

A grade-level **Planning Guide** provides an overview to your three modules and an introduction to three-dimensional teaching and learning.

The **Science Notebooks** chapter provides age-appropriate methods to support students in developing productive science notebooks. Access powerful research-based next-step strategies to maximize the effectiveness of the notebook as a formative assessment tool.

Science-Centered Language Development is a collection of standards-aligned strategies to support and enhance literacy development in the context of science—reading, writing, speaking, listening, and vocabulary development.

In **Taking FOSS Outdoors**, find guidance for managing the space, time, and materials needed to provide authentic, real-world learning experiences in students' local communities.

Teacher Resources also includes:

- Grade-level connections to Common Core ELA and Math standards
- Module-specific notebook, teacher, and assessment blackline masters.

Check FOSSweb for the latest updates to chapters in *Teacher Resources*.

3. FOSS Science Resources Student Book

The Teacher Toolkit includes one copy of the student book. Reading is an integral part of science learning. Reading informational text critically and effectively is an important component of today's ELA standards. Once students have engaged with phenomena firsthand, they go more in-depth with articles in *FOSS Science Resources*.

Articles from FOSS *Science Resources* complement and enhance the active investigations, giving students opportunities to:

- Ask and answer questions
- Use evidence to support their ideas
- Use text to acquire information
- Draw information from multiple sources
- Interpret illustrations to build understanding

Module includes Big Book

Interactive eBooks

FOSS Science Resources is available as a convenient, platform-neutral interactive student eBook with integrated audio, highlighted text, and links to videos and online activities. Student access to eBooks is available as an additional purchase.

Getting Started with Technology

FOSSweb.com

Easy access to program support resources

FOSSweb.com is your home for accessing the complete portfolio of digital resources in the FOSS program. Easily manage each of your modules, create class pages, and keep helpful references at your fingertips.

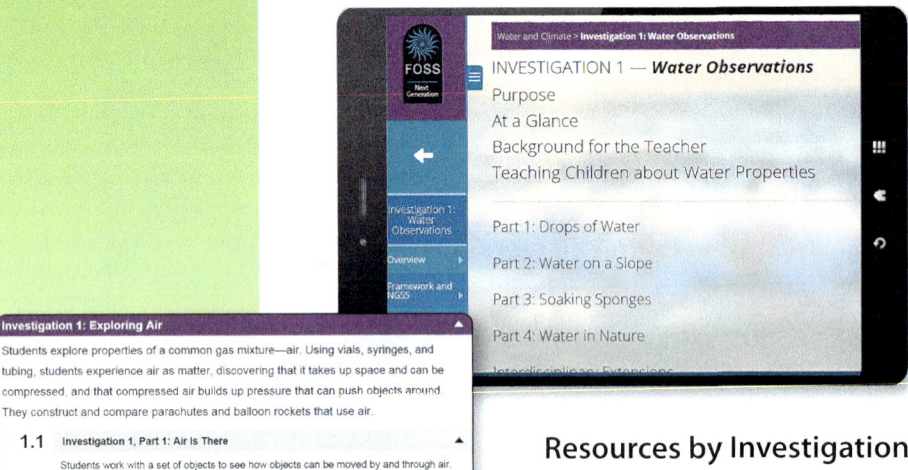

eInvestigations Guide

This easy-to-use interactive version of the *Investigations Guide* is mobile-friendly and offers simplified navigation, collapsible sections, and the ability to add customized notes.

Resources by Investigation

Easily access the duplication masters, online activities, and streaming videos needed for the current investigation part.

Teacher Preparation Videos

Videos provide helpful equipment setup instructions, safety information, and a summary of what students will do and learn throughout a part.

Interactive Whiteboard Lessons

Developed for SMART™ or Promethean boards, these resources help you facilitate each part of every investigation and give the class a visual reference.

Online Activities for Differentiating Instruction

FOSSweb digital resources provide engaging, interactive virtual investigations and tutorials that offer additional content and skill support for students. These experiences also help students who were absent catch up with class.

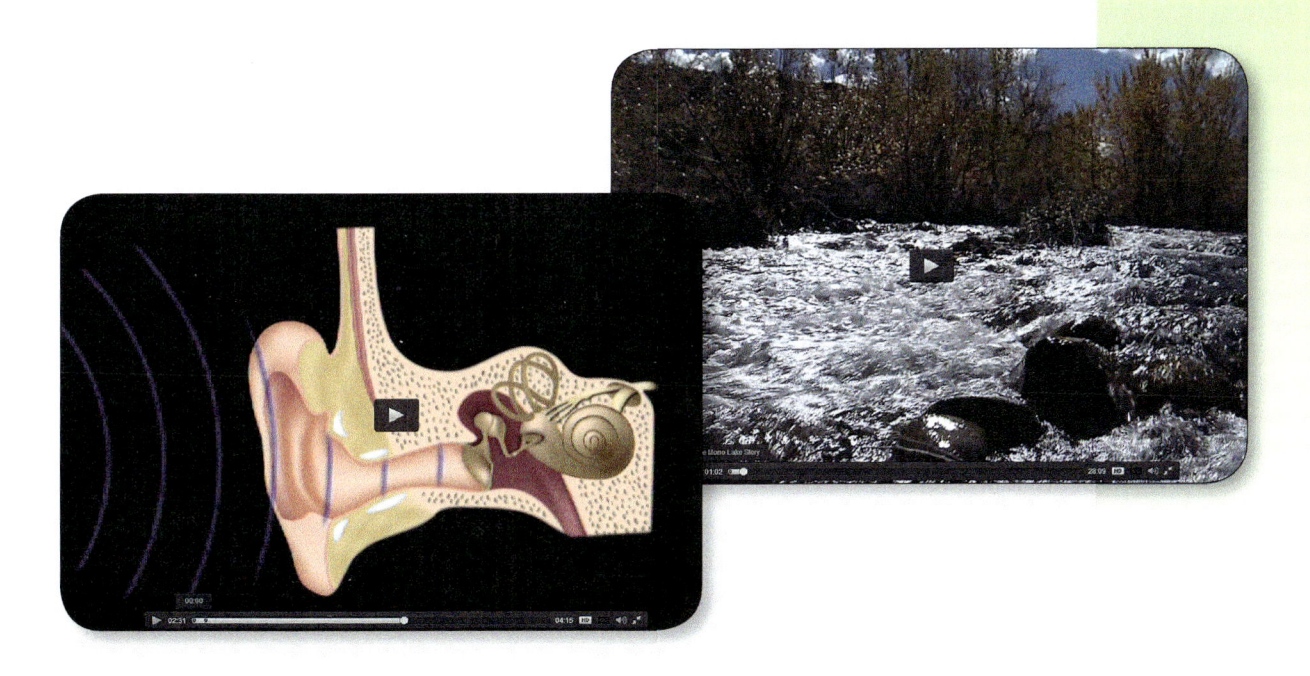

Streaming Videos

Videos are available on FOSSweb to support many investigations and often take students "on location" around the world or showcase experiments that would be too messy, expensive, or dangerous for the classroom.

Three-Dimensional Active Learning

The FOSS program has always placed student learning of science *practices* on equal footing with science *concepts and principles* and the NGSS and *Framework for K–12 Science Education* have provided a new language with which to articulate this. In each **FOSS Next Generation** investigation, students are engaged in the three dimensions of the NGSS to develop increasingly complex knowledge and understanding.

Science and engineering practices are the cognitive tools scientists and engineers use to answer questions and design solutions. FOSS students use these tools to gather evidence used to explain real-world phenomena.

Grade-level appropriate **disciplinary core ideas** are the concepts and established ideas of science. FOSS students develop these building blocks throughout investigations to make sense of phenomena.

Crosscutting concepts help students to connect the varied concepts and disciplines of science. FOSS students apply these concepts to different situations in order to make connections and develop comprehensive understanding.

FOSS Forward Thinking

The FOSS Vision

When the Full Option Science System (FOSS) began, the founders envisioned a science curriculum that was enjoyable, logical, and intuitive for teachers, and stimulating, provocative, and informative for students. Achieving this vision was informed by research in cognitive science, learning theory, and critical study of effective practice. The modular design of the FOSS product allowed users to select topics that aligned with district or state learning objectives, or simply resonated with their perception of comprehensive and reasonable science instruction. The original design of the FOSS Program was comprehensive in terms of coverage. FOSS was designed to provide real and meaningful student experience with important scientific ideas and to nurture developmentally appropriate knowledge of the objects, organisms, systems, and principles governing, the natural world.

The FOSS Next Generation Program

But the developers never envisioned FOSS to be a static curriculum, and now the Full Option Science System has evolved into a fully realized 21st century science program with authentic connection to the *Next Generation Science Standards (NGSS)*. The FOSS science curriculum is a comprehensive science program, featuring instructional guidance, student equipment, student reading materials, digital resources, and an embedded assessment system. The FOSS philosophy has always taken very seriously the teaching of good, comprehensive, accurate, science content using the methods of inquiry to advance that science knowledge. But the *Framework for K–12 Science Education*, on which the NGSS are based has allowed us to articulate our mission in a more coherent manner, using the vocabulary established by the authors of the *Framework*. The FOSS instructional design now strives to

a. communicate the disciplinary core ideas (content) of science, while

b. guiding and encouraging students to engage in or exercise the science and engineering practices (inquiry methods) to develop knowledge of the disciplinary core ideas, and

c. help students apprehend the crosscutting concepts (themes that unite core ideas, overarching concepts) that connect the learning experiences within a discipline and bridge meaningfully across disciplines as students gain more and more knowledge of the natural world.

> The Full Option Science System has evolved into a fully realized 21st century science program with authentic connection to the Next Generation Science Standards (NGSS).

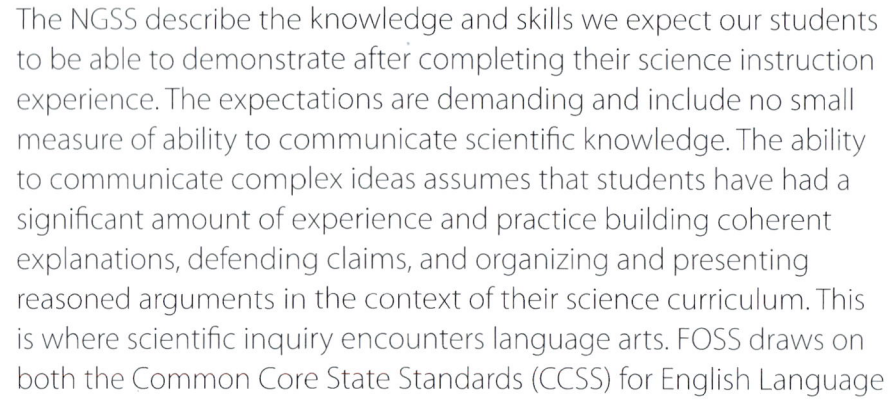

The NGSS describe the knowledge and skills we expect our students to be able to demonstrate after completing their science instruction experience. The expectations are demanding and include no small measure of ability to communicate scientific knowledge. The ability to communicate complex ideas assumes that students have had a significant amount of experience and practice building coherent explanations, defending claims, and organizing and presenting reasoned arguments in the context of their science curriculum. This is where scientific inquiry encounters language arts. FOSS draws on both the Common Core State Standards (CCSS) for English Language Arts and research data regarding the productive use of student science notebooks. FOSS developers realize that the most effective science program must seamlessly integrate science instruction goals and language arts skills. Science is one of the most engaging and productive arenas for introducing and exercising language arts skills: vocabulary, nonfiction (informational) reading, cause-and-effect relationships, on and on.

FOSS is strongly grounded in the realities of the classroom and the interests and experiences of the learners. The content in FOSS is teachable and learnable over multiple grade levels as students increase in their abilities to reason about and integrate complex ideas within and between disciplines.

FOSS is crafted with a structured, yet flexible, teaching philosophy that embraces the much-heralded 21st century skills; collaborative teamwork, critical thinking, and problem solving. The FOSS curriculum design promotes a classroom culture that allows both teachers and students to assume prominent roles in the management of the learning experience.

FOSS is built on the assumptions that understanding of core scientific knowledge and how science functions is essential for citizenship, that all teachers can teach science, and that all students can learn science. Formative assessment in FOSS creates a community of reflective practice. Teachers and students make up the community and establish norms of mutual support, trust, respect, and collaboration. The goal of the community is that everyone will demonstrate progress and will learn and grow.

ANIMALS TWO BY TWO — *Overview*

INTRODUCTION

The **Animals Two by Two Module** provides students with close and personal interaction with some common land and water animals. The animals and their survival needs are the engaging anchor phenomena. Students study the phenomena by observing and describing the structures of fish, birds, snails, earthworms, and isopods. Appropriate classroom habitats are established for some organisms and students find out what the animals need to live and grow. In four investigations, animals are studied in pairs. Students observe and care for one animal over time, and then they are introduced to another animal similar to the first but with differences in structure and behavior.

The guiding questions for the module are how are animal structures similar and different? and what do animals need to live and grow? The firsthand experiences are enriched with close-up photos of animals, some related to animals that students have observed in class and some to animals that are new. This process enhances observation, communication, and comparison.

Throughout the **Animals Two by Two Module**, students engage in science and engineering practices by asking questions, participating in collaborative investigations, observing, recording, and interpreting data to build explanations, and obtaining information from photographs. Students gain experiences that will contribute to an understanding of the crosscutting concepts of patterns; cause and effect; systems and system models; and structure and function.

Contents

Introduction	1
Module Matrix	2
FOSS Components	4
FOSS Instructional Design	8
Differentiated Instruction for Access and Equity	16
FOSS Investigation Organization	18
Establishing a Classroom Culture	20
Safety in the Classroom and Outdoors	24
Scheduling the Module	25
FOSS Contacts	26

The NGSS Performance Expectations bundled in this module include:

Life Sciences
K-LS1-1

Earth and Space Sciences
K-ESS2-2
K-ESS3-1

These performance expectations are also included in Trees and Weather.

▶ **NOTE**
The three modules for grade K in FOSS Next Generation are

Materials and Motion

Trees and Weather

Animals Two by Two

FOSS Full Option Science System

ANIMALS TWO BY TWO — Overview

	Investigation Summary	Guiding and Focus Questions for Phenomena
Inv. 1: Goldfish and Guppies	Students first engage with the phenomenon of fish. Students observe the structures and behaviors of goldfish. They feed the fish and enrich the environment in which the fish live. They compare the structures and behaviors of the goldfish to those of other fish, guppies. Students compare photos of fish and read about fish. Students then engage with the phenomenon of local birds. They go bird watching in the schoolyard and compare features and behaviors of birds.	*What do animals such as fish and birds need to live and grow?* What are the parts of a goldfish? What do goldfish need to live? What do goldfish do? How are guppies and goldfish different? How are they the same? What birds visit our schoolyard?
Inv. 2: Water and Land Snails	Students engage with the phenomenon of snails. Students observe the structures and behaviors of two kinds of water snails. Students work with a variety of seashells, discussing similarities and differences in their size, shape, color, and texture. Students match shell pairs, make designs, and create patterns. Students explore the schoolyard to find local land snails and compare their structures and behaviors to water snails.	*What do animals such as snails need to live and grow?* What are the parts of a water snail? How can shells be grouped? What do land snails do?
Inv. 3: Big and Little Worms	Students engage with the phenomenon of earthworms. Students dig for redworms, rinse them off, and look at their structures. They study their behavior. They construct worm jars and provide for the needs of the composting worms. Students observe how the worms change the plant material into soil. They compare the redworms to night crawlers, which are much larger. Students compare photos and read about worms and their activities in soil.	*What do animals such as worms need to live and grow?* What are the parts of a redworm? What do redworms need to live? How are redworms and night crawlers different? How are they the same?
Inv. 4: Pill Bugs and Sow Bugs	Students engage with the phenomenon of isopods. Students observe structures of two kinds of isopods. They learn to identify which are pill bugs and which are sow bugs. They hold isopod races. Students make a terrarium in which all the land animals live together. They compare photos and read about isopods. They read about and compare illustrations of a variety of animals and discuss the differences between living and nonliving things.	*What do animals such as isopods need to live and grow?* What are isopods? How are pill bugs and sow bugs different? How are they the same? How do isopods move? What do animals need to live?

Module Matrix

Content Related to Disciplinary Core Ideas	Reading/Technology	Assessment
• Fish are animals and have basic needs. • Fish have structures that help them live and grow. • Different kinds of fish have similar but different structures and behaviors. • Birds are animals that have basic needs. • Different kinds of birds have similar but different structures and behaviors.	**Science Resources Book** "Fish Same and Different" "Fish Live in Many Places" "Birds Outdoors" **Video** "The Urban Habitat of Peregrine Falcons" in *Is This a House for Hermit Crab?* (Extension)	**Embedded Assessment** Teacher observation **NGSS Performance Expectations** K-LS1-1 K-ESS2-2 K-ESS3-1
• Different kinds of snails have some structures and behaviors that are the same and some that are different. • Snails are animals and have basic needs—water, air, food, and space with shelter. • There is great diversity among snails. • Shells differ in size, shape, pattern, and texture. • Snails have senses.	**Science Resources Book** "Water and Land Snails" **Video** *Seashore Surprises*	**Embedded Assessment** Teacher observation **NGSS Performance Expectations** K-LS1-1 K-ESS2-2 K-ESS3-1
• Worms are animals and have basic needs. • Worms have identifiable structures. • Different kinds of worms have similar structures and behaviors; they also have differences (size, color). • Worm behavior is influenced by conditions in the environment. • Worms change plant material into soil.	**Science Resources Book** "Worms in Soil"	**Embedded Assessment** Teacher observation **NGSS Performance Expectations** K-LS1-1 K-ESS2-2 K-ESS3-1
• Isopods are animals and have basic needs—water, air, food, and space with shelter. • Different kinds of isopods have some structures and behaviors that are the same and some that are different. • There is great diversity among isopods. • Isopod behavior is influenced by conditions in the environment.	**Science Resources Book** "Isopods" "Animals All around Us" "Living and Nonliving" **Book** *Animals Two By Two* **Online Activity** "Find the Parent"	**Embedded Assessment** Teacher observation **NGSS Performance Expectations** K-LS1-1 K-ESS2-2 K-ESS3-1

ANIMALS TWO BY TWO — Overview

FOSS COMPONENTS

Teacher Toolkit for Each Module

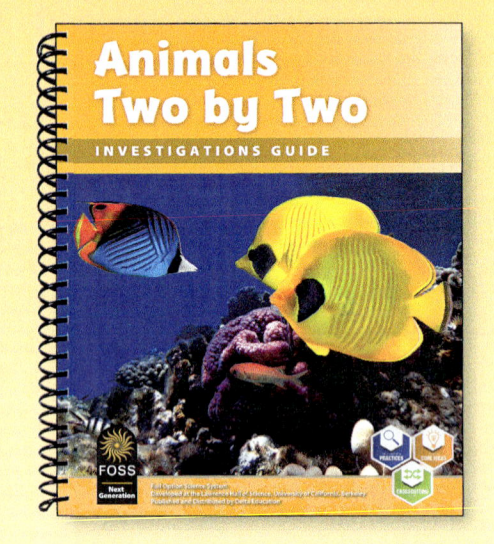

The FOSS Next Generation Program has three modules for kindergarten—Materials and Motion, Trees and Weather, and Animals Two by Two.

Each module comes with a *Teacher Toolkit* for that module. The *Teacher Toolkit* is the most important part of the FOSS Program. It is here that all the wisdom and experience contributed by hundreds of educators has been assembled. Everything we know about the content of the module, how to teach the subject, and the resources that will assist the effort are presented here. Each toolkit has three parts.

Investigations Guide. This spiral-bound document contains these chapters.

- Overview
- Framework and NGSS
- Materials
- Technology
- Investigations (four in this module)
- Assessment

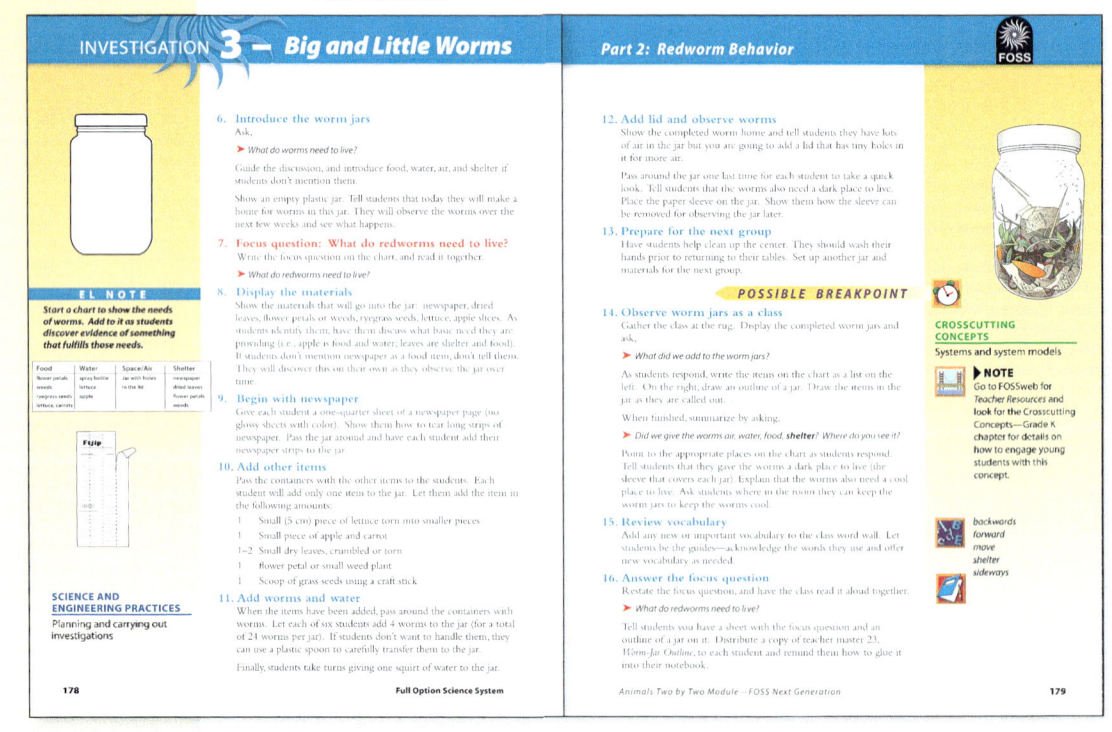

4 Full Option Science System

FOSS Components

FOSS Science Resources book. One copy of the student book of readings is included in the *Teacher Toolkit*.

Teacher Resources. These chapters can be downloaded from FOSSweb and are also in the bound *Teacher Resources* book.

- FOSS Program Goals
- Planning Guide—Grade K
- Science and Engineering Practices—Grade K
- Crosscutting Concepts—Grade K
- Sense-Making Discussions for Three-Dimensional Learning—Grade K
- Access and Equity
- Science Notebooks in Grades K–2
- Science-Centered Language Development
- FOSS and Common Core ELA—Grade K
- FOSS and Common Core Math—Grade K
- Taking FOSS Outdoors
- Teacher Masters
- Assessment Masters

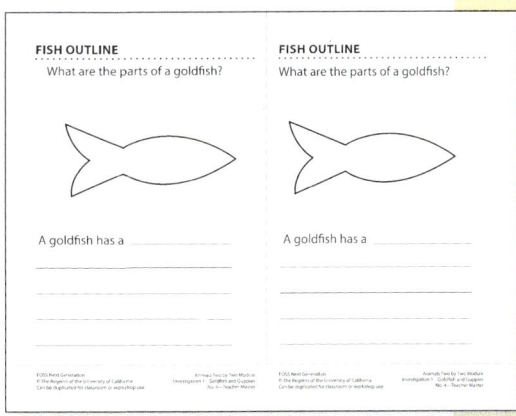

Equipment for Each Module or Grade Level

The FOSS Program provides the materials needed for the investigations in sturdy, front-opening drawer-and-sleeve cabinets. Inside, you will find high-quality materials packaged for a class of 32 students. Consumable materials are supplied for three uses before you need to resupply. Teachers may be asked to supply small quantities of common classroom materials.

Delta Education can assist you with materials management strategies for schools, districts, and regional consortia.

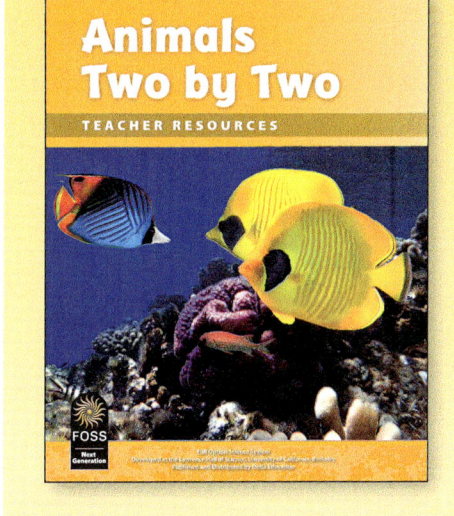

Animals Two by Two Module—FOSS Next Generation

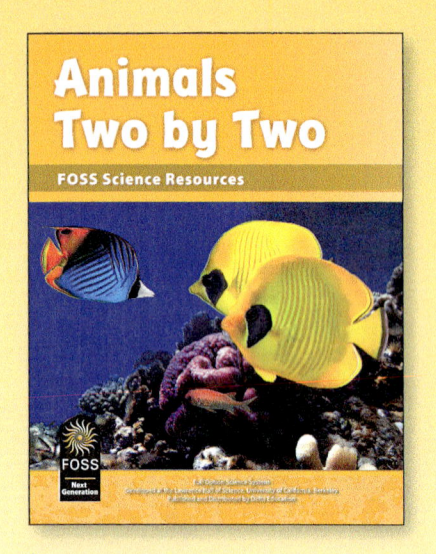

ANIMALS TWO BY TWO — Overview

FOSS Science Resources Books

FOSS Science Resources: Animals Two by Two is a book of original readings developed to accompany this module. The readings are referred to as articles in *Investigations Guide*. Students read the articles in the book as they progress through the module. The articles cover specific concepts, usually after the concepts have been introduced in the active investigation.

The articles in *Science Resources* and the discussion questions provided in *Investigations Guide* help students make connections to the science concepts introduced and explored during the active investigations. Concept development is most effective when students are allowed to experience organisms, objects, and phenomena firsthand before engaging the concepts in text. The text and illustrations help make connections between what students experience concretely and the ideas that explain their observations.

▶ **NOTE**
FOSS Science Resources: Animals Two by Two is also provided as a big book in the equipment kit.

Some snails live on land in **moist** places.

A garden is one place for land snails.

Full Option Science System

FOSS Components

Technology

The FOSS website opens new horizons for educators, students, and families, in the classroom or at home. Each module has digital resources for students and families—interactive simulations and online activities. For teachers, FOSSweb provides online teacher *Investigations Guides*; grade-level planning guides (with connections to ELA and math); materials management strategies; science teaching and professional development tools; contact information for the FOSS Program developers; and technical support. In addition FOSSweb provides digital access to PDF versions of the *Teacher Resources* component of the *Teacher Toolkit*, digital-only instructional resources that supplement the print and kit materials, and access to FOSSmap, the online assessment and reporting system for grades 3–8.

With an educator account, you can customize your homepage, set up easy access to the digital components of the modules you teach, and create class pages for your students with access to online activities.

▶ **NOTE**
To access all the teacher resources and to set up customized pages for using FOSS, log in to FOSSweb through an educator account. See the Technology chapter in this guide for more specifics.

Ongoing Professional Learning

The Lawrence Hall of Science and Delta Education strive to develop long-term partnerships with districts and teachers through thoughtful planning, effective implementation, and ongoing teacher support. FOSS has a strong network of consultants who have rich and experienced backgrounds in diverse educational settings using FOSS.

▶ **NOTE**
Look for professional development opportunities and online teaching resources on www.FOSSweb.com.

Animals Two by Two Module—FOSS Next Generation

ANIMALS TWO BY TWO — *Overview*

FOSS INSTRUCTIONAL DESIGN

FOSS is designed around active investigation that provides engagement with science concepts and science and engineering practices. Surrounding and supporting those firsthand investigations are a wide range of experiences that help build student understanding of core science concepts and deepen scientific habits of mind.

The Elements of the FOSS Instructional Design

- Using Formative Assessment
- Integrating Science Notebooks
- Active Investigation
- Taking FOSS Outdoors
- Engaging in Science-Centered Language Development
- Accessing Technology
- Reading *FOSS Science Resources* Books

Full Option Science System

FOSS Instructional Design

Each FOSS investigation follows a similar design to provide multiple exposures to science concepts. The design includes these pedagogies.

- Active investigation in collaborative groups: firsthand experiences with phenomena in the natural and designed worlds
- Recording in science notebooks to answer a focus question dealing with the scientific phenomenon under investigation
- Reading informational text in *FOSS Science Resources* books
- Online activities to acquire data or information or to elaborate and extend the investigation
- Outdoor experiences to collect data from the local environment or to apply knowledge
- Assessment to monitor progress and inform student learning

In practice, these components are seamlessly integrated into a curriculum designed to maximize every student's opportunity to learn.

A **learning cycle** employs an instructional model based on a constructivist perspective that calls on students to be actively involved in their own learning. The model systematically describes both teacher and learner behaviors in a coherent approach to science instruction.

A popular model describes a sequence of five phases of intellectual involvement known as the 5Es: engage, explore, explain, elaborate, and evaluate. The body of foundational knowledge that informs contemporary learning-cycle thinking has been incorporated seamlessly and invisibly into the FOSS curriculum design.

Engagement with real-world **phenomena** is at the heart of FOSS. In every part of every investigation, the investigative phenomenon is referenced implicitly in the focus question that guides instruction and frames the intellectual work. The focus question is a prominent part of each lesson and is called out for the teacher and student. The investigation Background for the Teacher section is organized by focus question—the teacher has the opportunity to read and reflect on the phenomenon in each part in preparing for the lesson. Students record the focus question in their science notebooks, and after exploring the phenomenon thoroughly, explain their thinking in words and drawings.

In science, a phenomenon is a natural occurrence, circumstance, or structure that is perceptible by the senses—an observable reality. Scientific phenomena are not necessarily phenomenal (although they may be)—most of the time they are pretty mundane and well within the everyday experience. What FOSS does to enact an effective engagement with the NGSS is thoughtful selection of scientific phenomena for students to investigate.

▶ **NOTE**
The anchor phenomena establish the storyline for the module. The investigative phenomena guide each investigation part. Related examples of everyday phenomena are incorporated into the readings, videos, discussions, formative assessments, outdoor experiences, and extensions.

Animals Two by Two Module—FOSS Next Generation

ANIMALS TWO BY TWO – Overview

Active Investigation

Active investigation is a master pedagogy. Embedded within active learning are a number of pedagogical elements and practices that keep active investigation vigorous and productive. The enterprise of active investigation includes

- context: sharing prior knowledge, questioning, and planning;
- activity: doing and observing;
- data management: recording, organizing, and processing;
- analysis: discussing and writing explanations.

Context: sharing, questioning, and planning. Active investigation requires focus. The context of an inquiry can be established with a focus question about a phenomenon or challenge from you or, in some cases, from students. (What do animals need to live and grow in a terrarium?) At other times, students are asked to plan a method for investigation. This might start with a teacher demonstration or presentation. Then you challenge students to plan an investigation, such as to find out what grows from the nodes of a potato. In either case, the field available for thought and interaction is limited. This clarification of context and purpose results in a more productive investigation.

Activity: doing and observing. In the practice of science, scientists put things together and take things apart, observe systems and interactions, and conduct experiments. This is the core of science—active, firsthand experience with objects, organisms, materials, and systems in the natural world. In FOSS, students engage in the same processes. Students often conduct investigations in collaborative groups of four, with each student taking a role to contribute to the effort.

The active investigations in FOSS are cohesive, and build on each other to lead students to a comprehensive understanding of concepts. Through investigations and readings, students gather meaningful data.

Data management: recording, organizing, and processing. Data accrue from observation, both direct (through the senses) and indirect (mediated by instrumentation). Data are the raw material from which scientific knowledge and meaning are synthesized. During and after work with materials, students record data in their science notebooks. Data recording is the first of several kinds of student writing.

Students then organize data so they will be easier to think about. Tables allow efficient comparison. Organizing data in a sequence (time) or series (size) can reveal patterns. Students process some data into graphs, providing visual display of numerical data. They also organize data and process them in the science notebook.

FOSS Instructional Design

Analysis: discussing and writing explanations. The most important part of an active investigation is extracting its meaning. This constructive process involves logic, discourse, and prior knowledge. Students share their explanations for phenomena, using evidence generated during the investigation to support their ideas. They conclude the active investigation by writing in their science notebooks a summary of their learning as well as questions raised during the activity.

Science Notebooks

Research and best practice have led FOSS to place more emphasis on the student science notebook. Keeping a notebook helps students organize their observations and data, process their data, and maintain a record of their learning for future reference. The process of writing about their science experiences and communicating their thinking is a powerful learning device for students. The science-notebook entries stand as credible and useful expressions of learning. The artifacts in the notebooks form one of the core exhibitions of the assessment system.

Full-size duplication masters are available on FOSSweb. Student work is entered partly in spaces provided on the notebook sheets and partly on adjacent blank sheets in the composition book. Look to the chapter in *Teacher Resources* called Science Notebooks in Grades K–2 for more details on how to use notebooks with FOSS.

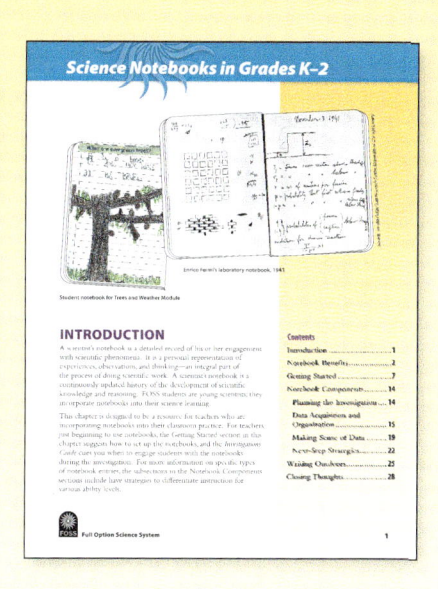

Animals Two by Two Module—FOSS Next Generation

ANIMALS TWO BY TWO – Overview

Reading in *FOSS Science Resources*

The *FOSS Science Resources* book, available in print and interactive eBooks, are primarily devoted to expository articles and biographical sketches. FOSS suggests that the reading be completed during language-arts time to connect to the Common Core State Standards for ELA. When language-arts skills and methods are embedded in content material that relates to the authentic experience students have had during the FOSS active learning sessions, students are interested, and they get more meaning from the text material.

Recommended strategies to engage students in reading, writing, speaking, and listening using the articles in the *FOSS Science Resources* books are included in the flow of Guiding the Investigation. In addition, a library of resources is described in the Science-Centered Language Development chapter in *Teacher Resources*.

The FOSS and Common Core ELA—Grade K chapter in *Teacher Resources* shows how FOSS provides opportunities to develop and exercise the Common Core State Standards for ELA practices through science. A detailed table identifies these opportunities in the three FOSS modules for kindergarten.

Engaging in Online Activities through FOSSweb

The simulations and online activities on FOSSweb are designed to support students' learning at specific times during instruction. Digital resources include streaming videos that can be viewed by the class or small groups. Resources can be used to review the active investigations and to support students who need more time with the concepts.

The Technology chapter provides details about the online activities for students and the tools and resources for teachers to support and enrich instruction. There are many ways for students to engage with the digital resources—in class as individuals, in small groups, or as a whole class, and at home with family and friends.

Full Option Science System

FOSS Instructional Design

Assessing Progress for Kindergarten

Assessment and teaching must be woven together to provide the greatest benefit to both the student and the teacher. Assessing young students is a process of planning what to assess, and observing, questioning, and recording information about student learning for future reference. Observing students as they engage in the activity and as they share notebook entries (drawings and words) reveals their thinking and problem-solving abilities. Questioning probes for understanding. Both observing and questioning will give you information about what individual students can and can't do, and what they know or don't know. This information allows you to plan your instruction thoughtfully. For example, if you find students need more experience comparing isopods, you can provide more time at a center for sorting and recording observations in their notebooks.

Use the techniques that work for you and your students and that fit with the overall kindergarten curriculum goals. The most detailed and reliable picture of students' growth emerges from information gathered by a variety of assessment strategies.

FOSS embedded assessments for kindergarten allow you and your students to monitor learning on a daily basis as you progress through the **Animals Two by Two Module**. You will find suggestions for what to assess in the Getting Ready section of each part of each investigation.

ANIMALS TWO BY TWO – Overview

Taking FOSS Outdoors

FOSS throws open the classroom door and proclaims the entire school campus to be the science classroom. The true value of science knowledge is its usefulness in the real world and not just in the classroom. Taking regular excursions into the immediate outdoor environment has many benefits. First of all, it provides opportunities for students to apply things they learned in the classroom to novel situations. When students are able to transfer knowledge of scientific principles to natural systems, they experience a sense of accomplishment.

In addition to transfer and application, students can learn things outdoors that they are not able to learn indoors. The most important object of inquiry outdoors is the outdoors itself. To today's youth, the outdoors is something to pass through as quickly as possible to get to the next human-managed place. For many, engagement with the outdoors and natural systems must be intentional, at least at first. With repeated visits to familiar outdoor learning environments, students may first develop comfort in the outdoors, and then a desire to embrace and understand natural systems.

The last part of most investigations is an outdoor experience. Venturing out will require courage the first time or two you mount an outdoor expedition. It will confuse students as they struggle to find the right behavior that is a compromise between classroom rigor and diligence and the freedom of recreation. With persistence, you will reap rewards. You will be pleased to see students' comportment develop into proper field-study habits, and you might be amazed by the transformation of students with behavior issues in the classroom who become your insightful observers and leaders in the schoolyard environment.

> **NOTE**
> The kit includes a set of four *Conservation* posters so you can discuss the importance of natural resources with students.

Teaching outdoors is the same as teaching indoors—except for the space. You need to manage the same four core elements of classroom teaching: time, space, materials, and students. Because of the different space, new management procedures are required. Students can get farther away. Materials have to be transported. The space has to be defined and honored. Time has to be budgeted for getting to, moving around in, and returning from the outdoor study site. All these and more issues and solutions are discussed in the Taking FOSS Outdoors chapter in *Teacher Resources*.

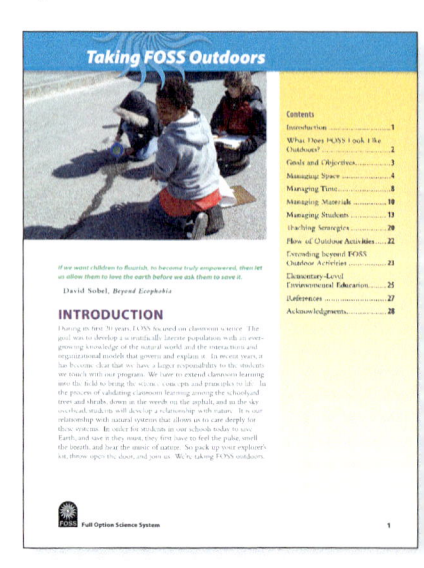

14 Full Option Science System

FOSS Instructional Design

Science-Centered Language Development and Common Core State Standards for ELA

The FOSS active investigations, science notebooks, *FOSS Science Resources* articles, and formative assessments provide rich contexts in which students develop and exercise thinking and communication. These elements are essential for effective instruction in both science and language arts—students experience the natural world in real and authentic ways and use language to inquire, process information, and communicate their thinking about scientific phenomena. FOSS refers to this development of language process and skills within the context of science as science-centered language development.

In the Science-Centered Language Development chapter in *Teacher Resources*, we explore the intersection of science and language and the implications for effective science teaching and language development. Language plays two crucial roles in science learning: (1) it facilitates the communication of conceptual and procedural knowledge, questions, and propositions, and (2) it mediates thinking—a process necessary for understanding. For students, language development is intimately involved in their learning about the natural world. Science provides a real and engaging context for developing literacy and language-arts skills identified in contemporary standards for English language arts.

The most effective integration depends on the type of investigation, the experience of students, the language skills and needs of students, and the language objectives that you deem important at the time. The Science-Centered Language Development chapter is a library of resources and strategies for you to use. The chapter describes how literacy strategies are integrated purposefully into the FOSS investigations, gives suggestions for additional literacy strategies that both enhance students' learning in science and develop or exercise English-language literacy skills, and develops science vocabulary with scaffolding strategies for supporting all learners. We identify effective practices in language-arts instruction that support science learning and examine how learning science content and engaging in science and engineering practices support language development.

Specific methods to make connections to the Common Core State Standards for English Language Arts are included in the flow of Guiding the Investigation. These recommended methods are linked to the CCSS ELA through ELA Connection notes. In addition, the FOSS and the Common Core ELA chapter in *Teacher Resources* summarizes all of the connections to each standard at the given grade level.

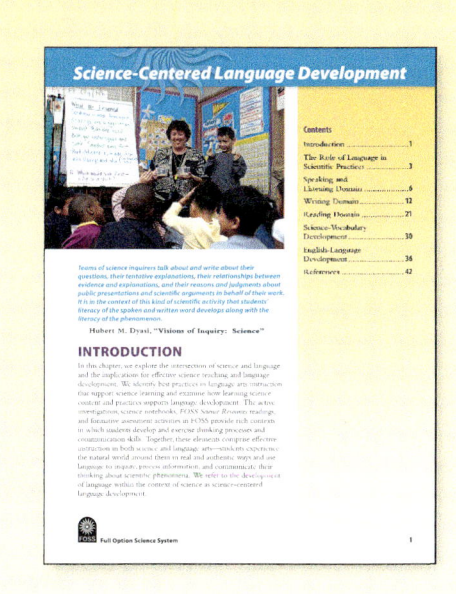

Animals Two by Two Module—FOSS Next Generation

ANIMALS TWO BY TWO — Overview

DIFFERENTIATED INSTRUCTION FOR ACCESS AND EQUITY

Learning from Experience

The roots of FOSS extend back to the mid-1970s and the Science Activities for the Visually Impaired and Science Enrichment for Learners with Physical Handicaps projects (SAVI/SELPH Program). As this special-education science program expanded into fully integrated (mainstreamed) settings in the 1980s, hands-on science proved to be a powerful medium for bringing all students together. The subject matter is universally interesting, and the joy and satisfaction of discovery are shared by everyone. Active science by itself provides part of the solution to full inclusion and provides many opportunities at the same time for differentiated instruction.

Many years later, FOSS began a collaboration with educators and researchers at the Center for Applied Special Technology (CAST), where principles of Universal Design for Learning (UDL) had been developed and applied. FOSS continues to learn from our colleagues about ways to use new media and technologies to improve instruction. Here are the UDL guiding principles.

Principle 1. Provide multiple means of representation. Give learners various ways to acquire information and demonstrate knowledge.

Principle 2. Provide multiple means of action and expression. Offer students alternatives for communicating what they know.

Principle 3. Provide multiple means of engagement. Help learners get interested, be challenged, and stay motivated.

FOSS for All Students

The FOSS Program has been designed to maximize the science learning opportunities for all students, including those who have traditionally not had access to or have not benefited from equitable science experiences—students with special needs, ethnically diverse learners, English learners, students living in poverty, girls, and advanced and gifted learners. FOSS is rooted in a 30-year tradition of multisensory science education and informed by recent research on UDL and culturally and linguistically responsive teaching and learning. Procedures found effective with students with special needs and students who are learning English are incorporated into the materials and strategies used with all students during the initial instruction phase. In addition, the **Access and Equity** chapter in *Teacher Resources* (or

"Active science by itself provides part of the solution to full inclusion and provides many opportunities at the same time for differentiated instruction."

Full Option Science System

Differentiated Instruction for Access and Equity

go to FOSSweb to download this chapter) provides strategies and suggestions for enhancing the science and engineering experiences for each of the specific groups noted above.

Throughout the FOSS investigations, students experience multiple ways of interacting with phenomena and expressing their understanding through a variety of modalities. Each student has multiple opportunities to demonstrate his or her strengths and needs, thoughts, and aspirations.

The challenge is then to provide appropriate follow-up experiences or enhancements appropriate for each student. For some students, this might mean more time with the active investigations or online activities. For other students, it might mean more experience and/or scaffolds for developing models, building explanations, or engaging in argument from evidence.

For some students, it might mean making vocabulary and language structures more explicit through new concrete experiences or through reading to students. It may help them identify and understand relationships and connections through graphic organizers. Interdisciplinary extensions in the arts, social studies, math, and language arts, as well as more advanced projects, are listed at the end of each investigation.

English Learners

The FOSS Program provides a rich laboratory for language development for English learners. A variety of techniques are provided to make science concepts clear and concrete, including modeling, visuals, and active investigations in small groups. Instruction is guided and scaffolded through carefully designed lesson plans, and students are supported throughout.

Science vocabulary and language structures are introduced in authentic contexts while students engage in hands-on learning and collaborative discussion. Strategies for helping all students read, write, speak, and listen are described in the Science-Centered Language Development chapter. A specific section on English learners provides suggestions for both integrating English language development (ELD) approaches during the investigation and for developing designated (targeted and strategic) ELD-focused lessons that support science learning.

Animals Two by Two Module—FOSS Next Generation

ANIMALS TWO BY TWO — Overview

FOCUS QUESTION
What are the parts of a goldfish?

SCIENCE AND ENGINEERING PRACTICES
Planning and carrying out investigations

DISCIPLINARY CORE IDEAS
LS1.C: Organization for matter and energy flow in organisms

CROSSCUTTING CONCEPTS
Patterns

TEACHING NOTE

This focus question can be answered with a simple yes or no, but the question has power when students support their answers with evidence. Their answers should take the form "Yes, because ____."

FOSS INVESTIGATION ORGANIZATION

Modules are subdivided into **investigations** (four in this module). Investigations are further subdivided into three to five **parts**. Each investigation has a general guiding question for the phenomenon students investigate, and each part of each investigation is driven by a specific **focus question**. The focus question, usually presented as the part begins, engages the student with the phenomenon and signals the challenge to be met, mystery to be solved, or principle to be uncovered. The focus question guides students' actions and thinking and makes the learning goal of each part explicit for teachers. Each part concludes with students recording an answer to the focus question in their notebooks.

The investigation is summarized for the teacher in the At-a-Glance chart at the beginning of each investigation.

Investigation-specific **scientific background** information for the teacher is presented in each investigation chapter, organized by the focus questions.

The **Teaching Children about** section makes direct connections to the NGSS foundation boxes for the grade level—Disciplinary Core Ideas, Science and Engineering Practices, and Crosscutting Concepts. This information is later presented in color-coded sidebar notes to identify specific places in the flow of the investigation where connections to the three dimensions of science learning appear. The Teaching Children about section ends with information about teaching and learning and a conceptual-flow graphic of the content.

The **Materials** and **Getting Ready** sections provide scheduling information and detail exactly how to prepare the materials and resources for conducting the investigation.

Teaching notes and **ELA Connections** appear in blue boxes in the sidebars. These notes comprise a second voice in the curriculum—an educative element. The first (traditional) voice is the message you deliver to students. The second (educative) voice, shared as a teaching note, is designed to help you understand the science content and pedagogical rationale at work behind the instructional scene. ELA Connections boxes provide connections to the Common Core State Standards for English Language Arts.

The **Getting Ready** and **Guiding the Investigation** sections have several features that are flagged in the sidebars. These include icons to remind you when a particular pedagogical method is suggested, as well as concise bits of information in several categories.

Full Option Science System

FOSS Investigation Organization

The **safety** icon alerts you to potential safety issues related to chemicals, allergic reactions, and the use of safety goggles.

The small-group **discussion** icon asks you to pause while students discuss data or construct explanations in their groups.

The **new-word** icon alerts you to a new vocabulary word or phrase that should be introduced thoughtfully.

The **vocabulary** icon indicates where students should review recently introduced vocabulary.

The **recording** icon points out where students should make a science-notebook entry.

The **reading** icon signals when the class should read a specific article in the *FOSS Science Resources* books.

The **technology** icon signals when the class should use a digital resource on FOSSweb.

The **assessment** icon appears when there is an opportunity to assess student progress by using embedded assessment.

The **crosscutting concepts** icon indicates an opportunity to expand on the concept by going to *Teacher Resources*, Crosscutting Concepts chapter.

The **outdoor** icon signals when to move the science learning experience into the schoolyard.

The **engineering** icon indicates opportunities for an experience incorporating engineering practices.

The **math** icon indicates an opportunity to engage in numerical data analysis and mathematics practice.

The **EL note** provides a specific strategy to assist English learners in developing science concepts.

To help with pacing, you will see icons for **breakpoints**. Some breakpoints are essential, and others are optional.

EL NOTE

POSSIBLE BREAKPOINT

Animals Two by Two Module—FOSS Next Generation

ANIMALS TWO BY TWO – Overview

ESTABLISHING A CLASSROOM CULTURE

Part of being a kindergartner is learning how to work collaboratively with others. However, students in primary grades are usually most comfortable working as individuals with materials. The abilities to share, take turns, and learn by contributing to a group goal are developing but are not reliable as learning strategies all the time. Because of this egocentrism and the need for many students to control materials or dominate actions, the FOSS kit includes a lot of materials. To effectively manage students and materials, FOSS offers some suggestions.

Small-Group Centers

Many of the kindergarten-level observations and investigations are conducted with small groups at a learning center. Limit the number of students at the center to six to ten at one time. When possible, each student will have his or her own equipment to work with. In some cases, students will have to share materials and equipment and make observations together. Primary students are good at working together independently.

As one group at a time is working at the center on a FOSS activity, other students will be doing something else. Over the course of an hour or more, plan to rotate all students through the center, or allow the center to be a free-choice station.

Whole-Class Discussions

Introducing and wrapping up the center activities require you to work for brief periods with the whole class. FOSS suggests for these introductions and wrap-ups that you gather the class at the rug or other location in the classroom where students can sit comfortably in a large group.

At the beginning of the year, explain and discuss norms for sense-making discussions. You might start by together making a class poster with visuals to represent what it looks like, sounds like, and feels like when everyone is working and learning together. Model discussion protocols that give all students opportunities to speak and listen, such as think-pair-share. Review the norms before sense-making discussions, and leave time for reflecting on how well the group adhered to the norms. More strategies for developing oral discourse skills can be found in Sense-Making Discussions for Three-Dimensional Learning and the Science-Centered Language Development chapters in *Teacher Resources* on FOSSweb.

This poster is an example of student responsibilities that the class discussed and adopted as their norms.

My Responsibilities
I agree that I will...
- explain my ideas.
- listen to others and show that I am listening.
- ask questions when I am confused or can't hear.
- connect my ideas to others' (explain, add to, respectfully disagree).
- participate because all ideas lead to learning (speak loud and clear).

Establishing a Classroom Culture

Collaborative Teaching and Learning

Collaborative learning requires a collective as well as individual growth mindset. A growth mindset is when people believe that their most basic abilities can be developed through dedication and hard work (see the research of Carol Dweck and her book *Mindset: The Psychology of Success*). As kindergartners learn to work together to make sense of phenomena and develop their inquiry and discourse skills, it's important to recognize and value their efforts to try new approaches, to share their ideas, and ask questions. Remind students that everyone in the classroom is a learner, and that learning happens when we try to figure things out. Here are a few ways to help students develop a growth mind-set for science and engineering.

- **Praise effort, not right answers.** When students are successful at a task, provide positive feedback about their level of engagement and effort in the practices, e.g., the efforts they put into careful observations, how well they reported their observations, the relevancy of their questions, how well they connected or applied new concepts, and their use of new vocabulary, etc. Also, try to provide feedback that encourages students to continue to improve their learning and exploring, e.g., is there another way you could try? Have you thought about _____? Why do you think _____?

- **Foster and validate divergent thinking.** During sense-making discussions, continually emphasize how important it is to share emerging ideas and to be open to the ideas of others in order to build understanding. Model for students how you refine and revise your thinking based on new information. Make it clear to students that the point is not for them to show they have the right answer, but rather to help each other arrive at new understandings. Point out positive examples of students expressing and revising their ideas. For example, Did you all notice how Carla changed her idea about _____?

Establishing a classroom culture that supports three-dimensional teaching and learning centers on collaboration. Helping students to work together in pairs and small groups, and to adhere to norms for discussions, are ways to foster collaboration. These structures along with the expectations that students will be negotiating meaning together as a community of learners, creates a learning environment where students are compelled to work, think, and communicate like scientists and engineers to help one another learn.

ANIMALS TWO BY TWO – *Overview*

Guides for Adult Helpers

On FOSSweb, you will find duplication masters for center instructions for some investigation parts. These sheets are intended as a quick reference for a family member or other adult who might be supervising the center and helping to guide the discussions. The sheets help that person keep the activity moving in a productive direction by suggesting questions and prompts to help students make sense of the phenomenon they are exploring. The sheets can be laminated or slipped into a clear-plastic sheet protector for durability.

When You Don't Have Adult Helpers

Some parts of investigations are designed for small groups, with an aide or a student's family member available to guide the activity and to encourage discussion and vocabulary development. We realize that there are many primary classrooms in which the teacher is the only adult present. Here are some ways to manage in that situation.

- Invite upper-elementary students to visit your class to help with the activities. They should be able to read the center instructions and conduct the activities with students. Remind older students to be guides and to let primary students do the activities themselves.

- Introduce each part of the activity with the whole class. Set up the center as described in *Investigations Guide*, but let students work at the center by themselves. Discussion might not be as rich, but most of the centers can be done independently by students once they have been introduced to the process. Be a 1-minute manager, checking on the center from time to time, offering a few words of advice or direction.

Establishing a Classroom Culture

Managing Materials

The Materials section lists the items in the equipment kit and any teacher-supplied materials. It also describes things to do to prepare a new kit and how to check and prepare the kit for your classroom. Individual photos of each piece of FOSS equipment are available for printing from FOSSweb, and can help students and you identify each item. (Photo equipment cards are available in English and Spanish formats.)

When Students Are Absent

When a student is absent for an activity, give him or her a chance to spend some time with the materials at a center. Another student might act as a peer tutor. Allow the student to bring home a *FOSS Science Resources* book to read with a family member.

ANIMALS TWO BY TWO – Overview

SAFETY IN THE CLASSROOM AND OUTDOORS

Following the procedures described in each investigation will make for a very safe experience in the classroom. You should also review your district safety guidelines and make sure that everything you do is consistent with those guidelines. Two posters are included in the kit: *Science Safety* for classroom use and *Outdoor Safety* for outdoor activities.

Look for the safety icon in the Getting Ready and Guiding the Investigation sections that will alert you to safety considerations throughout the module.

Safety Data Sheets (SDS) for materials used in the FOSS Program can be found on FOSSweb. If you have questions regarding any SDS, call Delta Education at 1-800-258-1302 (Monday–Friday, 8:00 a.m.–5:00 p.m. ET).

Science Safety in the Classroom

General classroom safety rules to share with students are listed here.

1. Listen carefully to your teacher's instructions. Follow all directions. Ask questions if you don't know what to do.
2. Tell your teacher if you have any allergies.
3. Never put any materials in your mouth. Do not taste anything unless your teacher tells you to do so.
4. Never smell any unknown material. If your teacher tells you to smell something, wave your hand over the material to bring the smell toward your nose.
5. Do not touch your face, mouth, ears, eyes, or nose while working with chemicals, plants, or animals.
6. Always protect your eyes. Wear safety goggles when necessary. Tell your teacher if you wear contact lenses.
7. Always wash your hands with soap and warm water after handling chemicals, plants, or animals.
8. Never mix any chemicals unless your teacher tells you to do so.
9. Report all spills, accidents, and injuries to your teacher.
10. Treat animals with respect, caution, and consideration.
11. Clean up your work space after each investigation.
12. Act responsibly during all science activities.

24 Full Option Science System

SCHEDULING THE MODULE

The Getting Ready section for each part of the investigation helps you prepare. It provides information on scheduling the investigation and introduces the tools and techniques used in the investigation. The first item in the Getting Ready section gives an estimated amount of time the part should take. A general rule of thumb is to plan 10 minutes to introduce the investigation to the whole class, about 15–20 minutes at the center for each group, about 10 minutes to wrap up the activity with the whole class, and a few minutes to transition to the groups. Notebook sessions can be done with the whole class after everyone has participated in the center activities. All of the outdoor sessions are whole-class activities. It will take about 8 weeks to complete the module.

Below is a list of the investigations and parts and the format of the investigation (whole class, center, or a combination of the two).

> **NOTE**
> The investigations are numbered, and we suggest that they be conducted in order since the concepts build from investigation to investigation.
>
> Be prepared—read the Getting Ready section thoroughly and review the teacher preparation video on FOSSweb.

INVESTIGATION	PART	ORGANIZATION
1. Goldfish and Guppies	1. The Structure of Goldfish	center
	2. Caring for Goldfish	center
	3. Goldfish Behavior	center
	4. Comparing Guppies to Goldfish	center/whole class
	5. Comparing Schoolyard Birds	whole class
2. Water and Land Snails	1. Observing Water Snails	center
	2. Shells	center/whole class
	3. Land Snails	center/whole class
3. Big and Little Worms	1. The Structure of Redworms	center
	2. Redworm Behavior	center
	3. Comparing Redworms to Night Crawlers	center/whole class
4. Pill Bugs and Sow Bugs	1. Isopod Observations	center
	2. Identifying Isopods	center/whole class
	3. Isopod Movement	center/whole class
	4. Animals Living Together	center/whole class

Animals Two by Two Module—FOSS Next Generation

ANIMALS TWO BY TWO — *Overview*

FOSS CONTACTS

General FOSS Program information

www.FOSSweb.com

www.DeltaEducation.com/FOSS

Developers at the Lawrence Hall of Science

FOSS@berkeley.edu

Customer service at Delta Education

www.DeltaEducation.com/contact.aspx

Phone: 1-800-258-1302, 8:00 a.m.–5:00 p.m. ET

FOSSmap (online component of FOSS assessment system)

http://FOSSmap.com/

FOSSweb account questions/access codes/help logging in

techsupport.science@schoolspecialty.com

Phone: 1-800-258-1302, 8:00 a.m.–5:00 p.m. ET

School Specialty online support

loginhelp@schoolspecialty.com

Phone: 1-800-513-2465, 8:30 a.m.–6:00 p.m. ET

FOSSweb tech support

support@FOSSweb.com

Professional development

www.FOSSweb.com/Professional-Development

Safety issues

www.DeltaEducation.com/SDS

Phone: 1-800-258-1302, 8:00 a.m.–5:00 p.m. ET

For chemical emergencies, contact Chemtrec 24 hours a day.

Phone: 1-800-424-9300

Sales and replacement parts

www.DeltaEducation.com/FOSS/buy

Phone: 1-800-338-5270, 8:00 a.m.–5:00 p.m. ET

ANIMALS TWO BY TWO – Framework and NGSS

INTRODUCTION TO PERFORMANCE EXPECTATIONS

"The NGSS are standards or goals, that reflect what a student should know and be able to do; they do not dictate the manner or methods by which the standards are taught.... Curriculum and assessment must be developed in a way that builds students' knowledge and ability toward the PEs [performance expectations]" (*Next Generation Science Standards*, 2013, page xiv).

This chapter shows how the NGSS Performance Expectations were bundled in the **Animals Two by Two Module** to provide a coherent set of instructional materials for teaching and learning.

This chapter also provides details about how this FOSS module fits into the matrix of the FOSS Program (page 33). Each FOSS module K–5 and middle school course 6–8 has a functional role in the FOSS conceptual frameworks that were developed based on a decade of research on science education and the influence of *A Framework for K–12 Science Education* (2012) and *Next Generation Science Standards* (NGSS, 2013).

The FOSS curriculum provides a coherent vision of science teaching and learning in the three ways described by the NRC *Framework*. First, FOSS is designed around learning as a developmental progression, providing experiences that allow students to continually build on their initial notions and develop more complex science and engineering knowledge. Students develop functional understanding over time by building on foundational elements (intermediate knowledge). That progression is detailed in the conceptual frameworks.

Second, FOSS limits the number of core ideas, choosing depth of knowledge over broad shallow coverage. Those core ideas are addressed at multiple grade levels in ever greater complexity. FOSS investigations at each grade level focus on elements of core ideas that are teachable and learnable at that grade level.

Third, FOSS investigations integrate engagement with scientific ideas (content) and the practices of science and engineering by providing firsthand experiences.

Teach the module with the confidence that the developers have carefully considered the latest research and have integrated into each investigation the three dimensions of the *Framework* and NGSS, and have designed powerful connections to the Common Core State Standards for English Language Arts.

Contents

Introduction to Performance Expectations 27

FOSS Conceptual Framework 32

Background for the Conceptual Framework in Animals Two by Two 34

Connections to NGSS by Investigation 38

FOSS Next Generation K–8 Scope and Sequence 42

The NGSS Performance Expectations bundled in this module include

Life Sciences
K-LS1-1

Earth and Space Sciences
K-ESS2-2
K-ESS3-1

The FOSS Trees and Weather Module also addresses these three performance expectations with a focus on plants.

Full Option Science System

ANIMALS TWO BY TWO — Framework and NGSS

Disciplinary Core Ideas Addressed

The **Animals Two by Two Module** connects with the NRC *Framework* for the grades K–2 grade band and the NGSS performance expectations for kindergarten. The module focuses on core ideas for life and earth sciences.

Life Sciences

Framework core idea LS1: From molecules to organisms: structures and processes—How do organisms live, grow, respond to their environment, and reproduce?

- **LS1.A: Structure and function**
  How do the structures of organisms enable life's functions? [All organisms have external parts. Different animals use their body parts in different ways to see, hear, grasp objects, protect themselves, move from place to place, and see, find, and take in food, water, and air. Plants also have different parts that help them survive, grow, and produce more plants.]

- **LS1.C: Organization for matter and energy flow in organisms**
  How do organisms obtain and use the matter and energy they need to live and grow? [All animals need food in order to live and grow. They obtain their food from plants or from other animals. Plants need water and light to live and grow.]

The following NGSS kindergarten performance expectation for LS1 is derived from the Framework disciplinary core ideas above.

- **K-LS1-1.** Use observations to describe patterns of what plants and animals (including humans) need to survive. [Clarification Statement: Examples of patterns could include that animals need to take in food but plants do not; the different kinds of food needed by different types of animals; the requirement of plants to have light; and that all living things need water.] The **FOSS Trees and Weather Module** addresses this performance expectation with a focus on plants.

DISCIPLINARY CORE IDEAS

A Framework for K–12 Science Education has four core ideas in life sciences.

LS1: From molecules to organisms: Structures and processes

LS2: Ecosystems: Interactions, energy, and dynamics

LS3: Heredity: Inheritance and variation of traits

LS4: Biological evolution: Unity and diversity

The questions and descriptions of the core ideas in the text on these pages are taken from the NRC *Framework* for the grades K–2 grade band to keep the core ideas in a rich and useful context.

The performance expectations related to each core idea are taken from the NGSS for kindergarten.

Introduction to Performance Expectations

Earth and Space Sciences

Framework core idea ESS2: Earth's systems—How and why is Earth constantly changing?

- **ESS2.E: Biogeology**
  How do living organisms alter Earth's processes and structures? [Plants and animals (including humans) depend on the land, water, and air to live and grow. They in turn can change their environment (e.g., the shape of land, the flow of water).]

The following NGSS kindergarten performance expectation for ESS2 is derived from the Framework disciplinary core idea above.

- **K-ESS2-2.** Construct an argument supported by evidence for how plants and animals (including humans) can change the environment to meet their needs. [Clarification Statement: Examples of plants and animals changing their environment could include a squirrel digs in the ground to hide its food and tree roots can break concrete.] The **FOSS Trees and Weather Module** addresses this performance expectation with a focus on plants.

Framework core idea ESS3: Earth and human activity—How do Earth's surface processes and human activities affect each other?

- **ESS3.A: Natural resources**
  How do humans depend on Earth's resources? [Living things need water, air, and resources from the land, and they try to live in places that have the things they need. Humans use natural resources for everything they do: for example, they use soil and water to grow food, wood to burn to provide heat or to build shelters, and materials such as iron or copper extracted from Earth to make cooking pans.]

The following NGSS kindergarten performance expectation for ESS3 is derived from the Framework disciplinary core idea above.

- **K-ESS3-1.** Use a model to represent the relationship between the needs of different plants or animals (including humans) and the places they live. [Clarification Statement: Examples of relationships could include that deer eat buds and leaves, therefore, they usually live in forested areas; and grasses need sunlight, so they often grow in meadows. Plants, animals, and their surroundings make up a system.] The **FOSS Trees and Weather Module** addresses this performance expectation with a focus on plants.

DISCIPLINARY CORE IDEAS

A Framework for K–12 Science Education has three core ideas in Earth and space sciences.

ESS1: Earth's place in the universe

ESS2: Earth's systems

ESS3: Earth and human activity

Two of these core ideas are in the NGSS performance expectations for kindergarten.

The questions and descriptions of the core ideas in the text on these pages are taken from the NRC *Framework* for the grades K–2 grade band to keep the core ideas in a rich and useful context.

The performance expectations related to each core idea are primarily taken from the NGSS for grades K–2.

▶ REFERENCES

National Research Council. *A Framework for K–12 Science Education: Practices, Crosscutting Concepts, and Core Ideas.* Washington, DC: National Academies Press, 2012.

NGSS Lead States. *Next Generation Science Standards: For States, by States.* Washington, DC: National Academies Press, 2013.

National Governors Association Center for Best Practices and Council of Chief State School Officers. *Common Core State Standards for English Language Arts and Literacy in History/Social Studies, Science, and Technical Subjects*, Washington, DC: 2010.

ANIMALS TWO BY TWO — Framework and NGS

SCIENCE AND ENGINEERING PRACTICES

A Framework for K–12 Science Education (National Research Council, 2012) describes eight science and engineering practices as essential elements of a K–12 science and engineering curriculum. Seven practices are incorporated into the learning experiences in the **Animals Two by Two Module**.

The learning progression for this dimension of the framework is addressed in *Next Generation Science Standards* 2013, volume 2, appendix F. Elements of the learning progression for practices recommended for kindergarten as described in the performance expectations appear in bullets below each practice.

Science and Engineering Practices Addressed

1. ***Asking questions***
   - Ask questions based on observations to find more information about the natural and/or designed worlds.

2. ***Developing and using models***
   - Use a model to represent relationships in the natural world.

3. ***Planning and carrying out investigations***
   - With guidance, plan and conduct an investigation collaboratively with peers.
   - Make observations (firsthand or from media) and/or measurements to collect data that can be used to make comparisons.
   - Make predictions based on prior experiences.

4. ***Analyzing and interpreting data***
   - Record information (observations, thoughts, and ideas).
   - Use and share pictures, drawings, and/or writings of observations.
   - Use observations (firsthand or from media) to describe patterns in the natural and designed world(s) in order to answer scientific questions.
   - Compare predictions (based on prior experiences) to what occurred (observable events).

5. ***Constructing explanations***
   - Make observations (firsthand or from media) to construct an evidence-based account for natural phenomena.

6. ***Engaging in argument from evidence***
   - Construct an argument with evidence to support a claim.

7. ***Obtaining, evaluating, and communicating information***
   - Read grade-appropriate texts and/or use media to obtain scientific information to describe patterns in the natural world.
   - Communicate information or solutions with others in oral and/or written forms using models and/or drawings that provide detail about scientific ideas.

Introduction to Performance Expectations

Crosscutting Concepts Addressed

Patterns
- Patterns in the natural and human-designed world can be observed, used to describe phenomena, and used as evidence.

Cause and effect
- Events have causes that generate observable patterns.

Systems and system models
- Objects and organisms can be described in terms of their parts.
- Systems in the natural and designed world have parts that work together.

Structure and function
- The shape and stability of structures of natural and designed objects are related to their function(s).

Connections: Understandings about the Nature of Science

Scientific knowledge is based on empirical evidence.
- Scientists look for patterns and order when making observations about the world.

Scientific investigations use a variety of methods.
- Scientific investigations begin with a question.
- Scientists use different ways to study the world.

CROSSCUTTING CONCEPTS

A Framework for K–12 Science Education describes seven crosscutting concepts as essential elements of a K–12 science and engineering curriculum. The crosscutting concepts listed here are those recommended for kindergarten in the NGSS, and are incorporated into the learning opportunities in the **Animals Two by Two Module**.

The learning progression for this dimension of the framework is addressed in *Next Generation Science Standards*, 2013, volume 2, appendix G, in the NGSS. Elements of the learning progression for crosscutting concepts recommended for kindergarten as described in the performance expectations, appear after bullets below each concept.

CONNECTIONS

See *Next Generation Science Standards*, 2013, volume 2, appendix H and appendix J, for more on these connections.

For details on learning connections to Common Core State Standards for English Language Arts and Math, see the chapters FOSS and Common Core ELA—Grade K and FOSS and Common Core Math—Grade K in *Teacher Resources*.

ANIMALS TWO BY TWO – Framework and NGS

FOSS CONCEPTUAL FRAMEWORK

In the last half decade, teaching and learning research has focused on learning progressions. The idea behind a learning progression is that **core ideas** in science are complex and wide-reaching, requiring years to develop fully—ideas such as the structure of matter or the relationship between the structure and function of organisms. From the age of awareness throughout life, matter and organisms are important to us. There are things students can and should understand about these core ideas in primary school years, and progressively more complex and sophisticated things they should know as they gain experience and develop cognitive abilities. When we as educators can determine those logical progressions, we can develop meaningful and effective curricula for students.

FOSS has elaborated learning progressions for core ideas in science for kindergarten through grade 8. Developing a learning progression involves identifying successively more sophisticated ways of thinking about a core idea over multiple years.

If mastery of a core idea in a science discipline is the ultimate educational destination, then well-designed learning progressions provide a map of the routes that can be taken to reach that destination. . . . Because learning progressions extend over multiple years, they can prompt educators to consider how topics are presented at each grade level so that they build on prior understanding and can support increasingly sophisticated learning. (National Research Council, *A Framework for K–12 Science Education*, 2012, page 26)

TEACHING NOTE

FOSS has conceptual structure at the module and strand levels. The concepts are carefully selected and organized in a sequence that makes sense to students when presented as intended.

The FOSS modules are organized into three domains: physical science, earth science, and life science. Each domain is divided into two strands, as shown in the table "FOSS Next Generation—K–8 Sequence." Each strand represents a core idea in science and has a conceptual framework.

- Physical Science: matter; energy and change
- Earth and Space Science: dynamic atmosphere; rocks and landforms
- Life Science: structure and function; complex systems

The sequence in each strand relates to the core ideas described in the NRC *Framework*. Modules at the bottom of the table form the foundation in the primary grades. The core ideas develop in complexity as you proceed up the columns.

FOSS Conceptual Framework

Information about the FOSS learning progression appears in the **conceptual framework** (page 35), which shows the structure of scientific knowledge taught and assessed in this module, and the **content sequence** (pages 36–37), a graphic and narrative description that puts this single module into a K–8 strand progression.

FOSS is a research-based curriculum designed around the core ideas described in the NRC *Framework*. The FOSS module sequence provides opportunities for students to develop understanding over time by building on foundational elements or intermediate knowledge leading to the understanding of core ideas. Students develop this understanding by engaging in appropriate science and engineering practices and exposure to crosscutting concepts. The FOSS conceptual frameworks therefore are *more detailed* and *finer grained* than the set of goals described by the NGSS performance expectations (PEs). The following statement reinforces the difference between the standards as a blueprint for assessment and a curriculum, such as FOSS.

Some reviewers of both public drafts [of NGSS] *requested that the standards specify the intermediate knowledge necessary for scaffolding toward eventual student outcomes. However, the NGSS are a set of goals. They are PEs for the end of instruction—not a curriculum. Many different methods and examples could be used to help support student understanding of the DCIs and science and engineering practices, and the writers did not want to prescribe any curriculum or constrain any instruction. It is therefore outside the scope of the standards to specify intermediate knowledge and instructional steps.* (Next Generation Science Standards, 2013, volume 2, page 342)

FOSS Next Generation—K–8 Sequence

	PHYSICAL SCIENCE		EARTH SCIENCE		LIFE SCIENCE	
	MATTER	ENERGY AND CHANGE	ATMOSPHERE AND EARTH	ROCKS AND LANDFORMS	STRUCTURE/ FUNCTION	COMPLEX SYSTEMS
6–8	Chemical Interactions	Waves; Gravity and Kinetic Energy; Electromagnetic Force; Variables and Design	Planetary Science; Earth History; Weather and Water		Heredity and Adaptation; Human Systems Interactions; Populations and Ecosystems; Diversity of Life	
5	Mixtures and Solutions		Earth and Sun		Living Systems	
4		Energy		Soils, Rocks, and Landforms	Environments	
3	Motion and Matter		Water and Climate		Structures of Life	
2	Solids and Liquids			Pebbles, Sand, and Silt	Insects and Plants	
1		Sound and Light	Air and Weather		Plants and Animals	
K	Materials and Motion		Trees and Weather		Animals Two by Two	

Animals Two by Two Module—FOSS Next Generation

ANIMALS TWO BY TWO — Framework and NGS

BACKGROUND FOR THE CONCEPTUAL FRAMEWORK
in Animals Two by Two

The year in kindergarten might be the first time many children are introduced to animals other than the familiar neighborhood animals that humans have invited into their homes as pets. The typical kindergartner's concept of animal is narrow, embracing for the most part a selection of mammals. When asked to recall the names of a few animals, kindergartners will provide lists that read like farm and zoo inventories. Cats, dogs, bears, birds, and the parade of other animals are not all animals to early-childhood students—they are cats, dogs, bears, and birds, each in a category of its own. The conceptual organization used by kindergartners does not yet recognize the superordinate set called animals that includes all the members in the kingdom of animals. This module will start expanding students' concept of animal by providing experiences with a number of animals that look and act quite unlike the barnyard model.

Life has proliferated on our planet for the last 3.5 billion years. There were no animals for most of that time—bacteria held a monopoly for the first 2 billion years. Animals probably emerged on the scene about 700 million years ago in forms similar to present-day sponges and jellyfish. An organism in the animal kingdom is multicellular and must eat to survive. Unlike members of the plant kingdom, animals cannot make their own food by photosynthesis (or a related process), so they must eat other organisms to get the energy needed to sustain life.

Kingdom Animalia has more members than any of the other kingdoms (Monera, Protista, Plantae, and Fungi). By some estimates 10 million different kinds of animals are living today, but many experts agree that the actual number could be many times greater. Better estimates are available for the number of animals that have backbones—mammals, fish, birds, reptiles, and amphibians. There are about 42,500 kinds of vertebrates. The great multitudes that make up the rest of the animal kingdom are known collectively as invertebrates, including the mollusks (clams, snails), crustaceans (crabs, shrimps, isopods), annelids (worms), and insects. Far and away the most populous class of animals is Insecta, with more members than all the other described animals combined.

In this module, students observe firsthand and describe aquatic vertebrates (goldfish and guppies), mollusks (water snails and land snails), annelids (redworms and night crawlers), and crustaceans (isopods).

Full Option Science System

FOSS Conceptual Framework

Through photos, they are introduced to a variety of birds. Obvious in their absence are the insects, which because of their rearing needs we saved for another grade, with a whole module devoted to these diverse and interesting animals.

There are many wonderful children's fiction books that portray animals and plants with human characteristics—plants that can walk and talk, and animals that talk and wear human clothes. These fanciful stories, whether serious or humorous, capture the imagination of children. But it is important to help young students distinguish fact from fantasy.

Through their firsthand experiences in the FOSS modules **Trees and Weather** and **Animals Two by Two,** kindergarten students will develop a scientific understanding about the structures and behaviors of real plants and animals. Students will become good observers of organisms, and they will be able to communicate those observations through words and drawings. This scientific understanding of animals and plants will help students appreciate imaginary animals and plants they read about in fictional stories.

After reading a fictional story that gives plants or animals attributes they do not really have, take the time to ask students what was real and what was imaginary in the story. Have students compare their own firsthand experiences with plants and animals to those in the story.

CONCEPTUAL FRAMEWORK
Life Science, Focus on Structure and Function:
Animals Two by Two

Structures and Function

Concept A   All living things need food, water, a way to dispose of waste, and an environment in which they can live.

- Animals have identifiable structures and behaviors.

Concept B   Reproduction is essential to the continued existence of every kind of organism. Organisms have diverse life cycles.

- Adult animals and plants can have offspring.

Complex Systems

Concept A   Organisms and populations of organisms are dependent on their environmental interactions both with other living things and with nonliving factors.

- A habitat is a place where animals live and their needs are met. There are many different kinds of habitats.
- Animals need air (oxygen), water, food, and space with shelter.
- Animals obtain their food from plants or from other animals.
- Animals can change their environment.

Concept D   Biological evolution, the process by which all living things have evolved over many generations from common ancestors, explains both the unity and diversity of species.

- Living things need water, air, and resources from the land; they live in places that have the things that they need. They can survive only where their needs are met.

Animals Two by Two Module—FOSS Next Generation

ANIMALS TWO BY TWO — Framework and NGS

Life Science Content Sequence

This table shows the FOSS modules that inform the structure and function and complex systems strands. The supporting elements in these modules (somewhat abbreviated) are listed. The elements for the **Animals Two by Two Module** are expanded in the sequence.

Module or course	LIFE SCIENCE — Structure and Function	LIFE SCIENCE — Complex Systems
Diversity of Life (middle school)	• All living things are made of cells • Cells have the same needs and perform the same functions as more complex organisms. • All living things need food, water, a way to dispose of waste, and an environment in which they can live (macro and microlevel). • Plants reproduce in a variety of ways.	• Adaptations are structures or behaviors of organisms that enhance their chances to survive and reproduce in their environment. • Biodiversity is the wide range of existing life-forms that have adapted to the variety of conditions on Earth, from terrestrial to marine ecosystems.
Living Systems (grade 5)	• Food provides animals with the materials they need for body repair and growth and is digested to release the energy they need. • Reproduction is essential to the continued existence of every kind of organism. • Humans and other animals have systems made up of organs that are specialized for particular body functions. • Animals detect, process, and use information about their environment to survive.	• Organisms obtain gases, water, and minerals from the environment and release waste matter back into the environment. • Matter cycles between air and soil, and among plants, animals, and microbes as these organisms live and die. • Organisms are related in food webs. • Some organisms, such as fungi and bacteria, break down dead organisms, operating as decomposers.
Environments (grade 4)	• Plants and animals have structures and behaviors that function in growth, survival, and reproduction. • Producers make their own food. • Animals obtain food from eating plants or eating other animals.	• Organisms have ranges of tolerance for environmental factors as a result of their internal and external structures. • Organisms interact in feeding relationships in ecosystems (food chains and food webs). • Difference in individual characteristics may give individuals an advantage in surviving.
Structures of Life (grade 3)	• A seed is a living organism. • Plants and animals have structures that function in growth and survival. • Reproduction is essential to the continued existence of every kind of organism. • Organisms have diverse life cycles. • Behavior of animals is influenced by internal and external cues.	• Organisms are related in food chains. • Different organisms can live in different environments; adaptations allow them to survive in that environment. • Changes in an organism's habitat are sometimes beneficial, sometimes harmful. • Many characteristics of organisms are inherited from parents.
Insects and Plants (grade 2)	• Insects need air, food, water, and space including shelter, and different insects meet these needs in different ways. • Plants depend on the environment for water and light to grow • Reproduction is essential to the continued existence of every kind of organism.	• Bees and other insects help some plants by moving pollen from flower to flower. • Animals interact with plants using them as food. They also assist in plant reproduction. • There are many different kinds of living things and they exist in different habitats on land and in water.
Plants and Animals (grade 1)	• Plants and animals have structures that function in growth and survival. • Reproduction is essential to the continued existence of every kind of organism. • Plants and animals grow and change and have predictable stages. • Adult plants and animals can have offspring.	• Plants make their own food. Plants depend on air, water, nutrients, and light to grow. • Plants are very much, but not exactly, like their parents. • There are many kinds of habitats. • Living things can survive only where their needs are met.
Animals Two by Two (grade K)		
Trees and Weather (grade K)	• Trees are living plants and have structures. • Trees go through predictable stages.	• Living things can survive only when their needs are met.

36   Full Option Science System

FOSS Conceptual Framework

> **NOTE**
> See the Assessment chapter at the end of this *Investigations Guide* for more details on how the FOSS embedded assessment opportunities align to the conceptual frameworks and the learning progressions.

The NGSS Performance Expectations addressed in this module include

Life Sciences
K-LS1-1

Earth and Space Sciences
K-ESS2-2
K-ESS3-1

See pages 28–29 in this chapter for more details on the kindergarten NGSS Performance Expectations.

Structure and Function	Complex Systems
• Animals have identifiable structures and behaviors. • Animals have basic needs. Land animals need air, water, food, and space with shelter. Water animals need the appropriate kind of water, oxygen from the water, food, and space with shelter. • Adult animals and plants can have offspring.	• A habitat is a place where animals live and their needs are met. There are many different kinds of habitats. • Animals eat plants and other animals. • Living things can survive only where their needs are met. • Organisms can change their environment.

Animals Two by Two

ANIMALS TWO BY TWO — Framework and NGSS

CONNECTIONS TO NGSS BY INVESTIGATION

	Science and Engineering Practices	Connections to Common Core State Standards for EL
Inv. 1: Goldfish and Guppies	Asking questions Developing and using models Planning and carrying out investigations Analyzing and interpreting data Constructing explanations Obtaining, evaluating, and communicating information	RI 1: Ask and answer questions about key details. RI 2: Identify main topic and retell key details. RI 3: Describe the connection between two ideas. RI 4: Ask and answer questions about unknown words. RI 5: Identify the front cover, back cover, and title page of a book. RI 7: Describe the relationship between illustrations and the text. RI 10: Actively engage in group reading activities with purpose and understanding. W 5: Strengthen writing. W 8: Gather information to answer a question. SL 1: Participate in collaborative conversations. SL 2: Ask and answer questions about key details and request clarification. SL 3: Ask and answer questions to seek help, information, or to clarify. SL 4: Describe with details.
Inv. 2: Water and Land Snails	Asking questions Planning and carrying out investigations Analyzing and interpreting data Constructing explanations Engaging in argument from evidence Obtaining, evaluating, and communicating information	RI 1: Ask and answer questions about key details. RI 2: Identify main topic and retell key details. RI 3: Describe the connection between two ideas. RI 7: Describe the relationship between illustrations and the text. RI 9: Identify similarities in and differences between two texts on the same topic. RI 10: Actively engage in group reading activities with purpose and understanding. W 5: Strengthen writing. W 7: Participate in shared research and writing projects. W 8: Gather information to answer a question. SL 1: Participate in collaborative conversations. SL 3: Ask and answer questions to seek help, information, or to clarify. SL 4: Describe with details.

Connections to NGSS by Investigation

Disciplinary Core Ideas		Crosscutting Concepts
LS1.A: Structure and function • All organisms have external parts. Different animals use their body parts in different ways to see, hear, grasp objects, protect themselves, move from place to place, and seek, find, and take in food, water, and air. Plants also have different parts (roots, stems, leaves, flowers, fruits) that help them survive and grow. (**foundational**) **LS1.C: Organization for matter and energy flow in organisms** • All animals need food in order to live and grow. They obtain their food from plants or from other animals. Plants need water and light to live and grow. **(K-LS1-1)**	**ESS2.E: Biogeology** • Plants and animals can change their environment. **(K-ESS2-2)** **ESS3.A: Natural resources** • Living things need water, air, and resources from the land, and they live in places that have the things they need. Humans use natural resources for everything they do. **(K-ESS3-1)**	Patterns Cause and effect Systems and system models Structure and function
LS1.A: Structure and function • All organisms have external parts. Different animals use their body parts in different ways to see, hear, grasp objects, protect themselves, move from place to place, and seek, find, and take in food, water, and air. Plants also have different parts (roots, stems, leaves, flowers, fruits) that help them survive and grow. (**foundational**) **LS1.C: Organization for matter and energy flow in organisms** • All animals need food in order to live and grow. They obtain their food from plants or from other animals. Plants need water and light to live and grow. **(K-LS1-1)**	**ESS2.E: Biogeology** • Plants and animals can change their environment. **(K-ESS2-2)** **ESS3.A: Natural resources** • Living things need water, air, and resources from the land, and they live in places that have the things they need. Humans use natural resources for everything they do. **(K-ESS3-1)**	Patterns Cause and effect Systems and system models Structure and function

ANIMALS TWO BY TWO — Framework and NGSS

Science and Engineering Practices	Connections to Common Core State Standards for ELA
Inv. 3: Big and Little Worms Asking questions Developing and using models Planning and carrying out investigations Analyzing and interpreting data Constructing explanations Engaging in argument from evidence Obtaining, evaluating, and communicating information	RI 1: Ask and answer questions about key details. RI 2: Identify main topic and retell key details. RI 3: Describe the connection between two ideas. RI 4: Ask and answer questions about unknown words. RI 7: Describe the relationship between illustrations and the text. RI 9: Identify similarities in and differences between two texts on the same topic. RI 10: Actively engage in group reading activities with purpose and understanding. W 5: Strengthen writing. W 8: Gather information to answer a question. SL 1: Participate in collaborative conversations. SL 4: Describe with details.
Inv. 4: Pill Bugs and Sow Bugs Asking questions Planning and carrying out investigations Analyzing and interpreting data Constructing explanations Obtaining, evaluating, and communicating information	RI 1: Ask and answer questions about key details. RI 2 Identify main topic and retell key details. RI 4: Ask and answer questions about unknown words. RI 6: Name and define the role of the author and illustrator. RI 7: Describe the relationship between illustrations and the text. RI 8: Identify the reasons an author gives to support points. RI 9: Identify similarities in and differences between two texts on the same topic. RI 10: Actively engage in group reading activities with purpose and understanding. W 5: Strengthen writing. W 8: Gather information to answer a question. SL 1: Participate in collaborative conversations. SL 2: Ask and answer questions about key details and request clarification. L 5a: Sort objects into categories.

Connections to NGSS by Investigation

Disciplinary Core Ideas		Crosscutting Concepts
LS1.A: Structure and function • All organisms have external parts. Different animals use their body parts in different ways to see, hear, grasp objects, protect themselves, move from place to place, and seek, find, and take in food, water, and air. Plants also have different parts (roots, stems, leaves, flowers, fruits) that help them survive and grow. (**foundational**) **LS1.C: Organization for matter and energy flow in organisms** • All animals need food in order to live and grow. They obtain their food from plants or from other animals. Plants need water and light to live and grow. (**K-LS1-1**)	**ESS2.E: Biogeology** • Plants and animals can change their environment. (**K-ESS2-2**) **ESS3.A: Natural resources** • Living things need water, air, and resources from the land, and they live in places that have the things they need. Humans use natural resources for everything they do. (**K-ESS3-1**)	Patterns Cause and effect Systems and system models Structure and function
LS1.A: Structure and function • All organisms have external parts. Different animals use their body parts in different ways to see, hear, grasp objects, protect themselves, move from place to place, and seek, find, and take in food, water, and air. Plants also have different parts (roots, stems, leaves, flowers, fruits) that help them survive and grow. (**foundational**) **LS1.C: Organization for matter and energy flow in organisms** • All animals need food in order to live and grow. They obtain their food from plants or from other animals. Plants need water and light to live and grow. (**K-LS1-1**)	**ESS2.E: Biogeology** • Plants and animals can change their environment. (**K-ESS2-2**) **ESS3.A: Natural resources** • Living things need water, air, and resources from the land, and they live in places that have the things they need. Humans use natural resources for everything they do. (**K-ESS3-1**)	Patterns Cause and effect Systems and system models Structure and function

ANIMALS TWO BY TWO — Framework and NGS

FOSS NEXT GENERATION K–8 SCOPE AND SEQUENCE

Grade	Physical Science	Earth Science	Life Science
6–8	Waves* Gravity and Kinetic Energy*	Planetary Science	Heredity and Adaptation* Human Systems Interactions*
	Chemical Interactions	Earth History	Populations and Ecosystems
	Electromagnetic Force* Variables and Design*	Weather and Water	Diversity of Life
5	Mixtures and Solutions	Earth and Sun	Living Systems
4	Energy	Soils, Rocks, and Landforms	Environments
3	Motion and Matter	Water and Climate	Structures of Life
2	Solids and Liquids	Pebbles, Sand, and Silt	Insects and Plants
1	Sound and Light	Air and Weather	Plants and Animals
K	Materials and Motion	Trees and Weather	Animals Two by Two

* Half-length course

Full Option Science System

ANIMALS TWO BY TWO — *Materials*

Contents

Introduction	43
Kit Inventory List	44
Materials Supplied by the Teacher	46
Preparing a New Kit	48
Preparing the Kit for Your Classroom	49
Care, Reuse, and Recycling	54

INTRODUCTION

The Animals Two by Two kit contains

- *Teacher Toolkit: Animals Two by Two*
  - 1 *Investigations Guide: Animals Two by Two*
  - 1 *Teacher Resources: Animals Two by Two*
  - 1 *FOSS Science Resources: Animals Two by Two*
- *FOSS Science Resources: Animals Two by Two* (1 big book and class set of student books)
- Permanent equipment for one class of 32 students
- Consumable equipment for three classes of 32 students

A new kit contains enough consumable items for at least three classroom uses before you need to restock. Some of the FOSS early-childhood investigations take place at a science center for groups of six to ten students at a time. For whole-class activities, use a materials station for the class materials.

Individual photos of each piece of FOSS equipment are available online for printing. For updates to information on materials used in this module and access to the Safety Data Sheets (SDS), go to www.FOSSweb.com. Links to replacement-part lists and customer service are also available on FOSSweb.

▶ **NOTE**
To see how all of the materials in the module are set up and used, view the teacher preparation video on FOSSweb.

▶ **NOTE**
Delta Education Customer Service can be reached at 1-800-258-1302.

FOSS Full Option Science System

ANIMALS TWO BY TWO — Materials

KIT INVENTORY List

Drawer 1 of 2

Equipment Condition

★ The student books, if included in your purchase, are shipped separately.

▶ **NOTE**
The teacher toolkit is shipped separately. However, there is space in drawer 1 to store your toolkit.

Print Materials

1	*Teacher Toolkit: Animals Two by Two* (1 *Investigations Guide*, 1 *Teacher Resources*, and 1 *FOSS Science Resources: Animals Two by Two*)
1	*FOSS Science Resources: Animals Two by Two*, big book
1	Bird silhouette set (6 sparrows, 3 robins, 1 crow)
1	Book, *Animals Two by Two*
1	Poster set, *Conservation*, 4/set
2	Posters, *Science Safety* and *Outdoor Safety*

Items for Investigation 1

1	Collecting net
100	Self-stick notes ✪

Items for Investigation 2

16	Cardboard pieces, 10 × 30 cm ✪
400	Shells, mollusks
12	Vials, 7 dram

Items for Investigation 3

10	Craft sticks
5	Jars, plastic, with screw lids with holes, 2 L
1	Tray, white plastic

Items for Investigation 4

2	Race tracks, laminated (each track comes in two parts)
12	Vials, 12 dram

Consumable Items

1	Fish food, 1.23 oz.
1	Seeds, ryegrass, package, 4 oz.
1	Soil, bag, 2 kg/bag

✪ These items might occasionally need replacement.

Full Option Science System

Drawer 2 of 2

Equipment Condition

Shared Items

6	Basins, clear plastic, 6 L	
6	Basin covers, plastic	
5	Bug boxes	
9	Containers, plastic, 1/2 L	
5	Container lids	
30	Cups, plastic, 250 mL (9 oz.)	
36	Cup lids	
1	Dechlorinator, bottle ✪	
1	Gravel, bag, 1kg/bag	
16	Hand lenses, three-power	
1	Plastic wrap, roll, clear ✪	
10	Spoons, plastic	
1	Spray mister	
4	Tunnels, square plastic pipe	

▶ **NOTE**
This module includes access to FOSSweb, which includes the streaming videos and online activities used throughout the module.

✪ These items might occasionally need replacement.

ANIMALS TWO BY TWO — Materials

> **NOTE**
> Throughout the *Investigations Guide*, we refer to materials not provided in the kit as "teacher-supplied." These materials are generally common or consumable items that schools and/or classrooms already have, such as rulers, paper towels, and computers. If your school/classroom does not have these items, they can be provided by teachers, schools, districts, or materials centers (if applicable). You can also borrow the items from other departments or classrooms, or request these items as community donations.

MATERIALS Supplied by the Teacher

Each part of each investigation has a Materials section that describes the materials required for that part. It lists materials needed for each student or group of students and for the class.

Be aware that you must supply some items. Each of these items is indicated in the materials list for each part of the investigation with an asterisk (★). Here is a summary list of those items by investigation.

For most investigations
- Chart paper and marking pen
- Drawing utensils (crayons, pencils, colored pencils, marking pens)
- Glue sticks
- Paper towels
- 1 Pitcher or water container
- Science notebooks (composition books)
- 1 Scissors
- Tape, transparent

For outdoor investigations
- Collecting bags for carrying materials
- Clipboards (optional)

Investigation 1: Goldfish and Guppies
- 1 Bird guide
- 1 Bunch of elodea (*Anacharis*), 6–8 sprigs
- Clipboard
- 1 Computer with Internet access
- 32 Envelopes (optional)
- 2 Goldfish
- 6 Guppies (4 females and 2 males)
- 1 Hole punch
- 1 Paper cutter (optional)
- 1 Projection system
- 10 Scissors
- 1 Piece of scratch paper
- 1 Stapler
- String, 27 m
- 64 Toilet paper tubes (empty)
- 3 Sheets of white paper
- Water, aged or treated

46   Full Option Science System

Investigation 2: Water and Land Snails

- Aged or treated water
- 1 Bunch of elodea (*Anacharis*)
- Chalk or egg shells
- 1 Computer with Internet access
- 12–15 Land snails (collected locally or within your state; see Step 2 of Preparing the Kit for Your Classroom on pages 49–50)
- Lettuce or carrot
- 10 Pieces of construction paper
- 1 Projection system
- 24 Water snails (pond snails and ramshorn snails)

Investigation 3: Big and Little Worms

- Flower petals or weeds and leaves
- Leaf litter
- Carrot, apple, or lettuce
- 5 Sheets of construction paper or file folders
- 5 Index cards (optional)
- 12 Night crawlers
- Oatmeal
- Packing tape, clear
- Paperclips, regular and jumbo
- Paper towels
- Pitcher
- Newspaper (no glossy sheets)
- 150 Redworms
- 8 Small objects, such as blocks, pencils, and centimeter cubes
- Water

Investigation 4: Pill Bugs and Sow Bugs

- Carrot, potato, lettuce
- 1–2 Flat rocks
- 25 Pill bugs
- Leaf litter
- Paper towels
- Scissors
- 1 Small garden plant
- 8 Small objects, such as blocks, pencils, and centimeter cubes
- 25 Sow bugs
- 1 Stopwatch or clock
- Tree bark
- Water

Animals Two by Two Module—FOSS Next Generation

ANIMALS TWO BY TWO — *Materials*

PREPARING *a New Kit*

If you choose to prepare the materials all at once with a group of volunteers, you can use the following guidelines for organization.

1. **Prepare the center instruction sheets**

   Each investigation part that involves a group of students at a center has a center instruction sheet written for a parent or other adult helper working with students. The sheet summarizes the information provided to the teacher in *Investigations Guide*. Use the teacher masters to print or make a copy of each of the center instructions, and either laminate the sheet or put it in a clear-plastic sheet protector. Take time to orient your adult volunteers or aides to the overall purposes of the activities and encourage them to facilitate but not direct student learning at the center. Below are the teacher master numbers for the center instruction sheets.

   No. 5   Center Instructions—The Structure of Goldfish
   No. 6   Center Instructions—Caring for Goldfish
   No. 7   Center Instructions—Goldfish Behavior
   No. 9   Center Instructions—Comparing Guppies to Goldfish
   No. 12 Center Instructions—Observing Water Snails
   No. 14 Center Instructions—Shells
   No. 15 Center Instructions—Land Snails
   No. 19 Center Instructions—The Structure of Redworms
   No. 21 Center Instructions—Redworm Behavior A
   No. 22 Center Instructions—Redworm Behavior B
   No. 24 Center Instructions—Comparing Redworms to Night Crawlers
   No. 26 Center Instructions—Isopod Observations
   No. 28 Center Instructions—Identifying Isopods
   No. 29 Center Instructions—Isopod Races
   No. 30 Center Instructions—Animals Living Together

PREPARING *the Kit for Your Classroom*

Some preparation is required each time you use the kit. Doing these things before beginning the module will make daily setup quicker and easier.

1. ## Inventory materials
   Before using a kit, conduct a quick inventory of all items in the kit. You can use the Kit Inventory List provided in this chapter to keep track of any items that are missing or in need of replacement. Information on ordering replacement items can be found at the end of this chapter. The kit contains enough consumables for at least three classes of 32 students.

2. ## Obtain organisms
   Most of the animals used in the module can be obtained from local sources. We encourage you to do this, especially for the fish and aquatic snails, as it will be more economical and ensure healthier organisms. Isopods are often best supplied from Delta Education; they will usually have pill bugs and sow bugs when you want them. Delta sells coupons for all the animals and plants used in the module. The coupons are available separately, so you could obtain some animals locally and order others. A coupon provides enough animals for one class. Because each coupon is redeemed separately, you receive the animals on the schedule that you plan. It is important that you plan ahead. Allow 4 weeks to receive your animals if you redeem the coupons by mail. You can also redeem the coupons by fax. You can fax your coupon to Delta at 1-800-282-1560. If you need any organisms in a hurry, call Delta at 1-800-258-1302 to expedite your order.

   Here are some tips for the most efficient ordering of organisms.

   - **Order early.** You can order months in advance; just specify when you want each organism to arrive. Make a note in your schedule book about their expected arrival.
   - **Schedule a midweek arrival of organisms.** This timing will prevent having a shipment sit in storage over a weekend.
   - **Let your school office personnel know when you expect the shipment.** It is likely that they will want to get those "bugs and snails" out of their office and to your room as soon as the package arrives.

Animals Two by Two Module—FOSS Next Generation

ANIMALS TWO BY TWO — Materials

- **Be prepared.** Read the Getting Ready and Background sections of each investigation for all the details on how to prepare for the arrival of your organisms and how to care for them.

Goldfish, guppies, and elodea (*Anacharis*). You can purchase fish and elodea at local pet stores. We encourage you to use local resources for these, especially for the fish, as they do not fare well during shipment. Work with your pet store to get healthy, inexpensive goldfish and guppies. It is our experience that feeder goldfish, although they are the cheapest, don't always survive in the classroom aquarium. You will need 1 bunch (6–8 sprigs) of elodea, two small goldfish, and six guppies. Ask your dealer to try to give you four female and two male guppies. If elodea is not available, ask for hornwort or another nonfloating water plant.

Water snails. You will need about 12–24 water snails of two kinds—pond snails and ramshorn snails. The ideal water snails for the classroom are the largest variety, on which you can easily observe the mouth and tentacles. The cheapest and easiest to obtain, however, are much smaller. If you have purchased elodea for your aquarium, you might already have some snails attached to the plant. You can obtain water snails from your local pet store or order them from Delta Education.

Land snails. You will need 20 or more land snails for the class: one for each student at a center, with a few more to replace any that get damaged. As part of the investigation, students collect local land snails from the schoolyard, or you provide them from a local garden.

If you do not have land snails in your community (in gardens or natural areas) and need to order them, find a biological supplier in your state that sells land snails. Land snails cannot be ordered through Delta Education, as they cannot be shipped across state lines.

Redworms and night crawlers. You can often obtain earthworms from local bait stores. Purchase about 150 redworms and 12 night crawlers. You might be able to dig worms from your own garden, or you can order them from Delta Education. Redworms are used for worm composting and are sometimes available at garden supply stores.

Pill bugs and sow bugs. Students will compare two kinds of isopods. You will need about 25 of each for Investigation 4. Although they can often be collected from local areas, ordering

▶ **NOTE**
The United States Department of Agriculture (USDA) controls the transport of land snails across state borders. If you live in the states of AZ, CA, NM, TX, or WA, you can obtain a USDA permit to allow you to receive shipments of the *Helix aspersa* land snail from outside your state. If you live in other states, it is not possible to obtain a permit. Go to FOSSweb for more information.

them from Delta Education will ensure a source of both kinds, regardless of the season—pill bugs (*Armadillidium*) and sow bugs (*Porcellio*).

3. **Plan for animal food**

   Each kind of animal will need a different kind of food. The Getting Ready section in each investigation provides all the details. Plan ahead to make sure you have the appropriate food when you need it.

   - Goldfish and guppies will eat tropical fish flakes, supplied in the kit. Fish will also munch on the elodea plants in the aquariums.
   - Land snails will eat bits of lettuce, carrots, and other vegetable matter. Pieces of white chalk added to the terrarium will provide snails with needed calcium.
   - Water snails will eat water plants and the algae that forms on the sides of the aquarium. They will also eat lettuce.
   - Worms will eat bits of lettuce, carrots, and other vegetable matter. Oatmeal and decaying leaves are also good food sources.
   - Isopods will survive well on slices of potato and carrot, and a moist paper towel.

4. **Identify animal containers**

   Large containers for habitats. Six 6 liter (L) clear plastic basins with lids serve as permanent and semipermanent housing for the animals. Use one basin for each kind of animal: guppies, land snails, and water snails. Start with two basins for the goldfish. After the water snails are added to the fish basin, one basin is free for earthworms in soil.

   Containers for distribution. The 1/2 L plastic containers with lids are used as collection containers or temporary housing for snails, worms, and isopods. It is fine to poke several small air holes in the lids.

5. **Check soil**

   Earthworms need soil. The kit provides potting soil, but loamy soil from a yard or garden will be even better. You can mix some local soil with the potting soil. You will need a little less than half a bag of soil (1 kilogram [kg]) for one class.

6. **Condition water for fish**

   Chlorine in regular tap water can be lethal to both goldfish and guppies. There are two ways to dechlorinate water. The first is to age the tap water by letting it sit in an open container for at least 24 hours. Chlorine that is dissolved in the water escapes into the air. Or add dechlorinating chemicals (included in the kit) to tap

Animals Two by Two Module—FOSS Next Generation

ANIMALS TWO BY TWO — Materials

water. In some water systems, chloramine, a newer additive, is used in place of chlorine, and it will not leave the water when exposed to air. You must use water-conditioner chemicals that specifically say "removes chloramine." Check at your local pet store or aquarium store to find out just what is recommended in your area to make sure the water is safe for the fish.

7. **Plan for the unexpected**
   With living organisms in the classroom, almost anything can happen. Stay flexible and deal with each situation as it comes up.

8. **Respect living animals**
   Plan to model the proper way to handle and care for animals. Work with students to develop their sense of responsibility for the well-being of animals in the classroom. Some students might not want to pick up a worm or a snail at first, but if you let them watch their peers handling the organism, most students will overcome the fearful responses they might have learned elsewhere.

9. **Plan for the end of the module**
   Plan for care of the animals at the end of the module. The organisms might find a permanent home in your classroom. That would be ideal, as you can continue informal observations for a longer time. You will need to provide containers for permanent habitats if the kit will be used by another teacher. Or you might pass the organisms on to the next user of the kit. Some students might be interested in taking the animals home (be sure to obtain their family's permission ahead of time). Or you can check with your district to see if there is a plan for reuse of FOSS organisms.

   If there is no other option for the organisms, euthanize them by placing them in the freezer for a day or two, and then dispose of them in the trash.

10. **Plan for science notebooks**
    See the Getting Ready section for Investigation 1, Part 1, for ways to organize the science notebooks for this module.

11. **Plan for the word wall and pocket chart**
    As the module progresses, you will add new vocabulary words to a word wall or pocket chart and model writing and responding to focus questions. See the Getting Ready section for Investigation 1, Part 1, for suggestions about how to do this in your classroom.

12. **Plan for focus-question charts**
    Each part of each investigation asks a focus question before and after the activity. You'll find these questions on teacher masters 2–3, *Focus Questions A* and *B*. Students will glue each focus

> **NOTE**
> Never release organisms in the local environment if they were not collected from that specific location.

EL NOTE
You might want to print out the FOSS equipment photo cards (from FOSSweb) to add to your word wall to help students with vocabulary.

question on a page in their science notebooks and respond to it with words or drawings. At the beginning of the module, you will need to scaffold the use of notebooks. Use a chart to model how to respond to the focus question in writing or drawings. See the Getting Ready section for Investigation 1, Part 1, for suggestions on how to do this in your classroom.

13. **Plan for letter home and home/school connections**

    Teacher master 1, *Letter to Family*, is a letter you can use to inform families about this module. The letter states the goals of the module and suggests some home experiences that can contribute to students' learning.

    There is a home/school connection for most investigations. Check the last page of each investigation for details, and plan when to print or make copies and send them home with students.

14. **Review indoor and outdoor safety rules**

    Early-childhood students should be allowed to demonstrate that they can act responsibly with materials, but they must be given guidelines for safe and appropriate use of materials. Work with students to develop those guidelines so they can participate in making behavioral rules and understand the rationale for the rules. Emphasize that materials do not go in mouths, ears, noses, or eyes. Encourage responsible actions toward other students.

    Two safety posters are included in the kit to post in the room—*Science Safety* and *Outdoor Safety*. The Getting Ready for Investigation 1, Part 1, offers suggestions for this discussion.

15. **Gather books from library**

    Check your local library for books related to this module. Visit FOSSweb for a list of appropriate trade books that relate to this module.

16. **Check FOSSweb for resources**

    Go to FOSSweb to review the print and digital resources available for this module, including the eGuide, eBook, Resources by Investigation, and *Teacher Resources*, including the grade-level Planning Guide.

▶ **NOTE**
The *Letter to Family* and *Home/School Connections* are available electronically on FOSSweb.

TEACHING NOTE

For a detailed discussion of methods for working effectively with students outdoors, see the Taking FOSS Outdoors chapter.

Animals Two by Two Module—FOSS Next Generation

ANIMALS TWO BY TWO — *Materials*

CARE, *Reuse, and Recycling*

When you finish teaching the module, inventory the kit carefully. Note the items that were used up, lost, or broken, and immediately arrange to replace the items. Use a photocopy of the Kit Inventory List in this chapter, and put your marks in the "Equipment Condition" column. Refill packages and replacement parts are available for FOSS by calling Delta Education at 1-800-258-1302 or by using the online replacement-part catalog (www.DeltaEducation.com).

Standard refill packages of consumable items are available from Delta Education. A refill package for a module includes sufficient quantities of all consumable materials (except those provided by the teacher) to use the kit with three classes of 32 students.

Here are a few tips on storing the equipment after use.

- Make sure items are clean and dry before storing them.
- When cleaning the basins, rinse them well with hot water to remove any residual soap.
- Make sure the posters and print materials are flat on the bottom of the box.

The items in the kit have been selected for their ease of use and durability. Make sure that items are clean and dry before putting them back in the kit. Small items should be inventoried (a good job for students under your supervision) and put into zip bags for storage. Any items that are no longer useful for science should be properly recycled.

ANIMALS TWO BY TWO – Technology

Contents

Introduction	55
Technology for Students	56
Technology for Teachers	58
Requirements for Accessing FOSSweb	62
Troubleshooting and Technical Support	64

INTRODUCTION

Technology is an integral part of the teaching and learning with FOSS Next Generation. FOSSweb is the Internet access to FOSS digital resources. FOSSweb gives students the opportunity to interact with simulations, images, and text—activities that enhance understanding of core ideas. It provides support for teachers, administrators, and families who are actively involved in implementing FOSS.

Different types of online activities are incorporated into investigations where appropriate. Each activity is marked with the technology icon in the *Investigations Guide*. You will sometimes show videos to the class. At other times, individuals or small groups of students will work online to review concepts or reinforce their understanding.

To use these digital resources, you should have at least one computer with Internet access that can be displayed to the class by an LCD projector with an interactive whiteboard or a large screen. Access to enough devices for students to work in small groups or one-on-one is recommended for other parts.

All FOSS online activities are available at www.FOSSweb.com for teachers, students, and families. We recommend you access FOSSweb well before starting the module, to set up your teacher-user account and to become familiar with the resources.

▶ **NOTE**
To get the most current information, download the latest Technology chapter on FOSSweb.

FOSS Full Option Science System

ANIMALS TWO BY TWO — *Technology*

TECHNOLOGY *for Students*

FOSS is committed to providing a rich, accessible technology experience for all FOSS students. Students access FOSSweb using a class login that you set up. Here are brief descriptions of selected resources for students on FOSSweb.

Online activities. The online simulations and activities are designed to support students' learning at all grades. They include virtual investigations and tutorials, grades 3–5, that review selected active investigations and support students who have difficulties with the materials or who have been absent. Summaries of some of the online activities are on the next page.

FOSS Science Resources—eBooks. The student book is available as an audio book on FOSSweb, accessible at school or at home. In addition, as premium content, *FOSS Science Resources* is available as an eBook on computer or tablet, either as a read-only PDF or in an interactive format that allows text to be read and provides points of interactivity. The eBook can also be projected for guided reading with the whole class.

Media library. A variety of media enhances students' learning and provides them with opportunities to obtain, evaluate, and communicate information. FOSS has reviewed print books and digital resources that are appropriate for students and prepared a list of these resources with links to content websites. There is also a list of regional resources for virtual and actual field trips for students to use in gathering information for projects, and a database of science and engineering careers. Other resources include vocabulary lists to promote use of academic language.

Home/school connections. Each module includes a letter to families, providing an overview of the goals and objectives of the module. There is also a Module Summary available for families to download. Most investigations have a home/school science activity that connects the classroom experiences with students' lives outside of school. These connections are available as PDFs on FOSSweb.

Class pages. Teachers with a FOSSweb account can easily set up class pages with notes and assignments for each class. Students and families can then access this class information online, using the teacher-assigned class login.

▶ **NOTE**
The following student-facing resources are available in Spanish on FOSSweb using a teacher's class page.

- Vocabulary
- Equipment photo cards
- eBooks
- Select streaming videos
- Home/school connections
- Audio books

Technology for Students

Animals Two by Two Online Activity

Here is the online activity used in the **Animals Two by Two Module**.

Investigation 4, Part 4: Animals Living Together

- **"Find the Parent"**
  Students match offspring to adult by comparing features that make the animals similar.

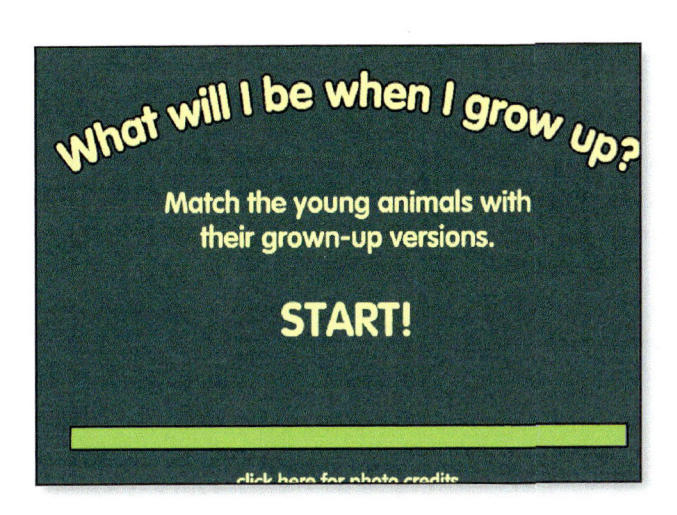

Animals Two by Two Module—FOSS Next Generation

57

ANIMALS TWO BY TWO — *Technology*

TECHNOLOGY *for Teachers*

The teacher side of FOSSweb provides access to all the student resources plus those designed for teaching FOSS. By creating a FOSSweb user account and activating your modules, you can personalize FOSSweb for easy access to your instructional materials. You can also set up a class login for students and their families.

Creating a FOSSweb Teacher Account

Setting up an account. Set up a teacher account on FOSSweb before you begin teaching a module. Go to FOSSweb and register for an account with your school e-mail address. Complete registration instructions are available online. If you have a problem, go to the Connecting with FOSS pull-down menu, and look at Technical Help and Access Codes. You can also access online tutorials for getting started with FOSSweb at www.FOSSweb.com/fossweb-walkthrough-videos.

Entering your access code. Once your account is set up, go to FOSSweb and log in. To gain access to all the teacher resources for your module, you will need to enter your access code. Your access code should be printed on the inside cover of your *Investigations Guide*. If you cannot find your FOSSweb access code, contact your school administrator, your district science coordinator, or the purchasing agent for your school or district.

Familiarize yourself with the layout of the site and the additional resources available when you log in to your account. From the module page, you will be able to access teacher masters, science notebook masters, assessment masters, the FOSSmap online assessment component, and other digital resources not available to "guests."

Explore the Resources by Investigation, as this will help you plan. This page makes it simple to select the investigation you are teaching, and view all the digital resources organized by part. Resources by Investigation provides immediate access to the streaming videos, online activities, science notebook masters, teacher masters, and other digital resources for each investigation part.

Setting up class pages and student accounts. To enable your students to log in to FOSSweb to see class assignments and student-facing digital resources, set up a class page and generate a username and password for the class. To do this, log in to FOSSweb and go to your teacher page. Under "My Class Pages," follow the instructions to create a new class page and to leave notes for students. Note: student access to the student eBook from your class page requires premium content.

▶ **NOTE**
For more information about FOSS premium content, including pricing and ordering, contact your local Delta sales representative by visiting www.DeltaEducation.com or by calling 1-800-258-1302.

Technology for Teachers

Support for Teaching FOSS

FOSSweb is designed to support teachers using FOSS. FOSSweb is your portal to instructional tools to make teaching efficient and effective. Here are some of the tools available to teachers.

- **Grade-level Planning Guide.** The Planning Guide provides strategies for three-dimensional teaching and learning.

- **Resources by Investigation.** The Resources by Investigation organizes in one place all the print and online instructional materials you need for each part of each investigation.

- **Investigations eGuide.** The eGuide is the complete *Investigations Guide* component of the *Teacher Toolkit*, in an electronic web-based format for computers or tablets. If your district rotates modules among several teachers, this option allows all teachers easy access to *Investigations Guide* at all times.

- **Teacher preparation videos.** Videos present information to help you prepare for a module, including detailed investigation information, equipment setup and use, safety, and what students do and learn in each part of the investigation.

- **Interactive whiteboard resources for grades K–5.** You can use these interactive files with or without an interactive whiteboard to facilitate each part of each investigation. You'll need to download the appropriate software to access the files. Links for software downloads are on FOSSweb.

- **Focus questions.** The focus questions address the phenomenon for each part of each investigation, and are formatted for classroom projection and for printing, so that students can glue each focus question into their science notebooks.

- **Module updates.** Important updates cover teacher materials, student equipment, and safety considerations.

- **Module teaching notes.** These notes include teaching suggestions and enhancements to the module, sent in by experienced FOSS users.

- **Home/school connections.** These masters include an introductory letter home (with ideas to reinforce the concepts being taught) and the home/school connection sheets.

- **State and regional resources.** Listings of resources for your geographic region are provided for virtual and actual field trips and for students to use as individual or class projects.

- **Access to FOSS developers.** Through FOSSweb, teachers have a connection to the FOSS developers and expert FOSS teachers.

▶ **NOTE**
There are two versions of the eGuide, a PDF-based eGuide that mimics the hard copy guide, and an HTML interactive eGuide that allows you to write instructional notes and to interface with online resources from the guide.

▶ **NOTE**
The following resources are available on FOSSweb in Spanish.

Teacher-facing resources:
- Teacher masters
- Assessment masters
- Focus questions
- Interactive whiteboard files

Student-facing resources:
- Vocabulary
- Equipment photo cards
- eBooks
- Select streaming videos
- Home/school connections
- Audio books

Animals Two by Two Module—FOSS Next Generation

ANIMALS TWO BY TWO — *Technology*

Technology for Differentiated Instruction

Some resources are for differentiated instruction. They can be used by students at home or by you as part of classroom instruction.

- **Online activities.** The online simulations and activities described earlier in this chapter are designed to support student learning and are often used during instruction. They include virtual investigations and student tutorials for grades 3–5 that you can use to support students who have difficulties with the materials or who have been absent. Tutorials require students to record data and answer concluding questions in their notebooks. In some cases, the notebook sheet used in the classroom investigation is also used for the virtual investigation.

- **Vocabulary.** The online word list has science-related vocabulary and definitions used in the module (in both English and Spanish).

- **Equipment photo cards.** Equipment cards provide labeled photos of equipment that students use in the investigations. Cards can be printed and posted on the word wall as part of instruction.

- **Student eBooks.** Student access to audio-only *FOSS Science Resource* books requires basic access. With premium content, students can access the books from any Internet-enabled device. The eBooks are available in PDF and interactive versions. The PDF version mimics the hard copy book. The interactive eBook reads the text to students—highlighting the text as it is read—and provides students with video clips and online activities.

- **Streaming videos used for extensions.** Some videos are part of the instruction in the investigation and are in Resources by Investigation for each part. Those videos also appear again in the digital resources under Streaming Videos along with other videos that extend concepts presented in a module.

- **Recommended books, websites, and careers database.** FOSS-recommended books, websites, and a Science and Engineering Careers Database that introduces students to a variety of career options and diversity of individuals engaged in those careers are provided.

- **Regional resources.** This list provides local resources that can be used to enhance instruction. The list includes website links and PDF documents from local sources.

▶ **NOTE**
The eBook is premium content for students.

Technology for Teachers

Support for Classroom Materials Management

- **Materials chapter.** A PDF of the Materials chapter in *Investigations Guide* is available to help you prepare for teaching. A list, organized by drawer, shows the materials included in the FOSS kit for a given module. You can print and use this list for inventory and to monitor equipment condition.

- **Safety Data Sheets (SDS).** A link takes you to the latest safety sheets, with information from materials manufacturers on the safe handling and disposal of materials.

- **Plant and animal care.** This section includes information on caring for organisms used in the investigations.

Professional Learning Connections

FOSSweb provides PDF files of professional development chapters, mostly from *Teacher Resources*, that explain how to integrate instruction to improve learning. Some of them are
- Sense-Making Discussions for Three-Dimensional Learning
- Science-Centered Language Development
- FOSS and Common Core English Language Arts and Math
- Access and Equity
- Taking FOSS Outdoors

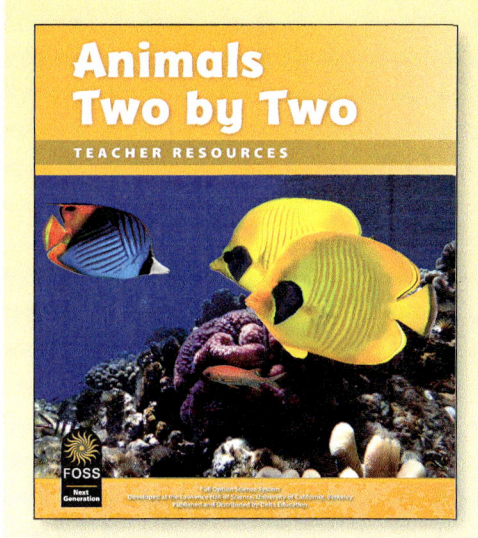

Animals Two by Two Module—FOSS Next Generation

ANIMALS TWO BY TWO – Technology

> **NOTE**
> It is strongly recommended that you visit FOSSweb to review the most recent minimum system requirements.

REQUIREMENTS *for Accessing* FOSSweb

FOSSweb Technical Requirements

To use FOSSweb, your computer must meet minimum system requirements and have a compatible browser and recent versions of Flash Player, QuickTime, and Adobe Reader. Many online activities have been updated to an HTML5 version compatible with all devices. (Those designated with "Flash" after the title require Flash Player.) The system requirements are subject to change. It is strongly recommended that you visit FOSSweb to review the most recent minimum system requirements and any plug-in requirements. There, you can access the "Tech Specs and Info" page to confirm that your browser has the minimum requirements to support the online activities.

Preparing your browser. FOSSweb requires a supported browser for Windows or Mac OS with a current version of the Flash Player plug-in, the QuickTime plug-in, and Adobe Reader or an equivalent PDF reader program. You may need administrator privileges on your computer in order to install the required programs and/or help from your school's technology coordinator.

By accessing the "Tech Specs and Info" page on FOSSweb, you can check compatibility for each computer you will use to access FOSSweb, including your classroom computer, computers in a school computer lab, and a home computer. The information on FOSSweb contains the most up-to-date technical requirements for all devices, including tablets and mobile devices.

Support for plug-ins and reader. Flash Player and Adobe Reader are available on www.adobe.com as free downloads. QuickTime is available for free from www.apple.com. FOSS does not support these programs. Please go to the program's website for troubleshooting information.

Requirements for Accessing FOSSweb

Other FOSSweb Considerations

Firewall or proxy settings. If your school has a firewall or proxy server, contact your IT administrator to add explicit exceptions in your proxy server and firewall for FOSSweb Akamai video servers. For more specific information on servers for firewalls, refer to "Tech Specs and Info" on FOSSweb.

Classroom technology setup. FOSS has a number of digital resources and makes every effort to accommodate users with different levels of access to technology. The digital resources can be used in a variety of ways and can be adapted to a number of classroom setups.

Teachers with classroom computers and an LCD projector, interactive whiteboard, or a large screen will be able to show online materials to the class. If you have access to a computer lab, or enough computers in your classroom for students to work in small groups, you can set up time for students to use the FOSSweb digital resources during the school day. Teachers who have access to only a single computer will find a variety of resources on FOSSweb that can be used to assist with teacher preparation and materials management.

Teachers who have tablets available for student use and have premium content can download the FOSS eBook app onto devices for easy student access to the FOSS eBooks. Instructions for downloading the app can be found on FOSSweb on the Module Detail Page for any module. You'll find them under the Digital-Only Resources section and then under the tab for Student eBooks.

Displaying online content. Throughout each module, you may occasionally want to project online components for instruction through your computer. To do this, you will need a computer with Internet access and either an LCD projector and a large screen, an interactive whiteboard, or a document camera arranged for the class to see.

You might want to display the notebook and teacher masters to the class. In Resources by Investigation, you'll have the option of downloading the masters to project or to copy. Choose "to project" if you plan on projecting to the class. These masters are optimized for a projection system. The "to copy" versions are sized to minimize paper use when photocopying for the class.

> **NOTE**
> FOSSweb activities are designed for a minimum screen size of 1024 × 768. It is recommended that you adjust your screen resolution to 1024 × 768 or higher.

ANIMALS TWO BY TWO — *Technology*

TROUBLESHOOTING *and Technical Support*

If you experience trouble with FOSSweb, you can troubleshoot in a variety of ways.

1. Test your browser to make sure you have the correct plug-in and browser versions. Even if you have the necessary plug-ins installed on your computer, they may not be recent enough to run FOSSweb correctly. Go to FOSSweb, and select the "Tech Specs and Info" page to review the most recent system requirements and check your browser.
2. Check the FAQs on FOSSweb for additional information that may help resolve the problem.
3. Empty the cache from your browser and/or quit and relaunch.
4. Restart your computer, and make sure all computer hardware turns on and is connected correctly.

If you are still experiencing problems after taking these steps, send FOSSweb Tech Support an e-mail to support@FOSSweb.com. In addition to describing the problem you are experiencing, include the following information about your computer: Mac or PC, operating system, browser name and version, plug-in names and versions. This will help troubleshoot the problem.

Where to Get Help

For further questions about FOSSweb, please don't hesitate to contact our technical support team.

Account questions/help logging in

School Specialty Online Support
techsupport.science@schoolspecialty.com
loginhelp@schoolspecialty.com

Phone: 1-800-513-2465, 8:30 a.m.–6:00 p.m. ET

General FOSSweb technical questions

FOSSweb Tech Support
support@FOSSweb.com

> **NOTE**
> The FOSS digital resources are available online on FOSSweb. You can always access the most up-to-date technology information, including help and troubleshooting, on FOSSweb.

INVESTIGATION 1 – *Goldfish and Guppies*

Part 1
The Structure of Goldfish 78

Part 2
Caring for Goldfish 86

Part 3
Goldfish Behavior 91

Part 4
Comparing Guppies to Goldfish 96

Part 5
Comparing Schoolyard Birds 104

Guiding question for phenomenon: What do animals such as fish and birds need to live and grow?

PURPOSE

Students have firsthand experiences with two related phenomena—goldfish and guppies. Through observation and discussion, students gather information about fish structures and behaviors and how those characteristics relate to the needs of the animals. Students apply their understandings to local outdoor birds.

Content

- Fish are animals and have basic needs—water with oxygen, food, and space with shelter.
- Fish have structures that help them live and grow—to find food, sense their habitat, and move from place to place.
- Different kinds of fish have similar but different structures and behaviors.
- Birds are animals that have basic needs.
- Different kinds of birds have similar but different structures and behaviors.

Practices

- Observe and compare the structures and behavior of two kinds of fish and ask questions based on observations.
- Observe and record changes in an aquarium over time.

Science and Engineering Practices

- Asking questions
- Developing and using models
- Planning and carrying out investigations
- Analyzing and interpreting data
- Constructing explanations
- Obtaining, evaluating, and communicating information

Disciplinary Core Ideas

LS1: How do organisms live, grow, respond to their environment, and reproduce?
LS1.A: Structure and function
LS1.C: Organization for matter and energy flow in organisms
ESS2: How and why is Earth constantly changing?
ESS2.E: Biogeology
ESS3: Earth and human activity
ESS3.A: Natural resources

Crosscutting Concepts

- Patterns
- Cause and effect
- Systems and system models
- Structure and function

FOSS Full Option Science System

INVESTIGATION 1 – *Goldfish and Guppies*

	Investigation Summary	Time	Focus Question for Phenomenon, Practices
PART 1	**The Structure of Goldfish** Students observe goldfish living in a simple aquarium. They look for and name different parts of the fish, such as fins, tail, mouth, and gills. They look to see if all the fish are alike, or if there are differences such as color and size. They draw a picture and dictate a sentence to record what they see.	**Introduction** 5 minutes **Center** 15–20 minutes **Notebook** 15 minutes	**What are the parts of a goldfish?** **Practices** Asking questions Planning and carrying out investigations Analyzing and interpreting data
PART 2	**Caring for Goldfish** Students learn how to care for goldfish, giving them food and fresh water, and adding plants to the aquarium. With each addition, students describe the fish behavior they observe.	**Introduction** 5 minutes **Center** 15–20 minutes **Notebook** 15 minutes	**What do goldfish need to live?** **Practices** Asking questions Analyzing and interpreting data
PART 3	**Goldfish Behavior** Students add a tunnel to the aquarium to observe how the fish respond. They make their own paper aquariums to model the fish behavior they have observed.	**Introduction** 5 minutes **Center** 15 minutes **Notebook** 15 minutes	**What do goldfish do?** **Practices** Developing and using models Analyzing and interpreting data
PART 4	**Comparing Guppies to Goldfish** Students compare the structures and behaviors of guppies to those of goldfish, and identify the guppies by gender.	**Center** 15 minutes **Notebook** 15 minutes **Reading** 30 minutes	**How are guppies and goldfish different? How are they the same?** **Practices** Asking questions Constructing explanations Obtaining, evaluating, and communicating information
PART 5	**Comparing Schoolyard Birds** Students go bird watching to observe and compare the structures and behaviors of two types of common schoolyard birds.	**Outdoors** 3 sessions **Notebook** 15 minutes **Reading** 15 minutes	**What birds visit our schoolyard?** **Practices** Asking questions Planning and carrying out investigations Analyzing and interpreting data Constructing explanations Obtaining, evaluating, and communicating information

At a Glance

Content Related to DCIs	Writing/Reading	Assessment
• Fish have structures that help them live and grow—to find food, sense their habitat, and move from place to place. • All animals deserve respect and gentle care.	**Science Notebook Entry** Draw or write words to answer the focus question.	**Embedded Assessment** Teacher observation
• Fish are animals and have basic needs—water with oxygen, food, and space with shelter. • Fish have structures that help them live and grow—to find food, sense their habitat, and move from place to place.	**Science Notebook Entry** Draw or write words to answer the focus question.	**Embedded Assessment** Teacher observation
• Fish have structures that help them live and grow—to find food, sense their habitat, and move from place to place.	**Science Notebook Entry** Draw or write words to answer the focus question.	**Embedded Assessment** Teacher observation
• Different kinds of fish have similar but different structures and behaviors.	**Science Notebook Entry** Draw or write words to answer the focus questions. **Science Resources Book** "Fish Same and Different" "Fish Live in Many Places"	**Embedded Assessment** Teacher observation
• Birds are animals that have basic needs. • Different kinds of birds have similar but different structures and behaviors.	**Science Notebook Entry** Bird Outline **Science Resources Book** "Birds Outdoors"	**Embedded Assessment** Teacher observation **NGSS Performance Expectations addressed in this investigation** K-LS1-1 K-ESS2-2 K-ESS3-1

Animals Two by Two Module—FOSS Next Generation

INVESTIGATION 1 – Goldfish and Guppies

> **TEACHING NOTE**
>
> *Refer to the grade-level Planning Guide chapter in* Teacher Resources *for a summary explanation of the phenomena students investigate in this module using a three-dimensional learning approach.*

BACKGROUND *for the Teacher*

The anchor phenomenon for this module is that animals of all kinds have various needs to live and grow. Each investigation provides students with firsthand experiences to care for and study the **structures** and behaviors of two examples of closely related organisms. Students' first experience is with animals that live in water.

Most evolutionary biologists think that life on Earth began in the ancient seas. Some organisms evolved to take up life on land; others found the aquatic environment totally suitable for life. The thousands of kinds of **fish** are examples of **animals** that, with few exceptions, are adapted for life in **water** and no place else.

Many organisms are known by common names that include the word *fish*, but are not true fish. Jellyfish, starfish, and a motley horde known collectively as shellfish are biologically quite **different** from the fish that are most familiar to young children. Bony fish have backbones, placing them in the subphylum of animals that have vertebrae. Unlike other vertebrates, however, fish live and breathe only underwater. It might appear that a fish in an **aquarium** is gasping for breath all the time, but in actuality it is filling its mouth and throat with water and then forcing the trapped water out through the gill openings on its sides, just past the **head**. Blood passing through the gills absorbs oxygen dissolved in the water, and circulates it to tissues throughout the body. Fish assume the temperature of their surroundings, making them ectothermic.

Fish have been around for more than 300 million years, and there are more kinds of fish than all other kinds of vertebrates put together. In fact, three out of every four vertebrates are fish. The bony fish comprise trout, catfish, tuna, cod, **goldfish**, and thousands of others. Add to these the cartilaginous fish—sharks and rays—and you have accounted for most of the fish that are living today.

Some 30,000 species of fish **swim** the seas, lakes, and streams of the world, exhibiting a staggering variety of sizes, shapes, and habits. One of the smallest fish is a tiny goby found in the Philippines—less than half the size of the typical **guppy**—and the largest fish is the whale shark, a gentle, plodding giant 12 meters (m) long and weighing 18 metric tons.

"Some 30,000 species of fish swim the seas, lakes, and streams of the world, exhibiting a staggering variety of sizes, shapes, and habits."

Background for the Teacher

What Are the Parts of a Goldfish?

The "typical" bony fish is streamlined to move through water, and equipped with a predictable array of structures: two lidless **eyes** mounted on the sides of the head, two nostrils for sensing the environment through smell, and a **mouth** in front with several rows of teeth. The teeth are used primarily for grasping prey, which are usually swallowed whole, rather than for ripping or chewing. Just behind the head are two bony **gill** covers that can open and close, and two pairs of **fins**—two pectorals (where arms would be) and two ventrals (where legs would be). A dorsal (back) fin, anal fin, and tail fin round out the complement of fins. Along each side of the fish runs a curved line that starts just **above** the gill cover and ends at the **tail**. This lateral line contains sensory organs that detect changes in pressure in the fish's environment. The whole body, with the exception of the face and the fins, may be covered with a set of overlapping **scales** to provide protective armor.

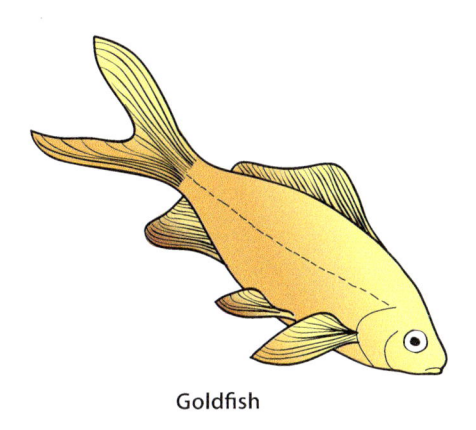

Goldfish

What Do Goldfish Need to Live?

Goldfish and guppies are among the hardiest aquarium fish and the easiest to care for. Students will enjoy hours of fun observing their behavior and caring for them as they become part of your classroom. Although they are easy to care for, there are some things that you should keep in mind to ensure success with your new aquarium. Suitable water, sufficient oxygen, correct temperature (within a fairly large range for these fish), and correct feeding are the most important things to consider.

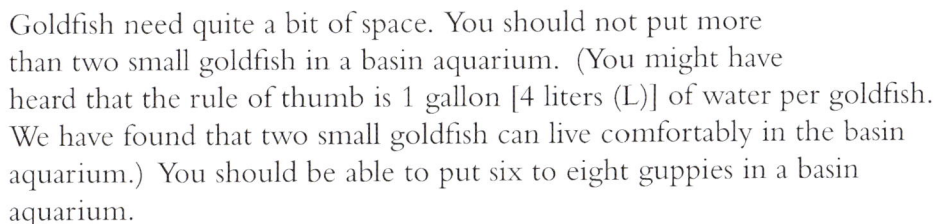

Goldfish need quite a bit of space. You should not put more than two small goldfish in a basin aquarium. (You might have heard that the rule of thumb is 1 gallon [4 liters (L)] of water per goldfish. We have found that two small goldfish can live comfortably in the basin aquarium.) You should be able to put six to eight guppies in a basin aquarium.

Chlorine in regular tap water can be lethal to both goldfish and guppies. There are two ways to dechlorinate water. The first is to age the tap water by letting it sit in an open container for at least 24 hours. Chlorine dissolved in the water escapes into the air. Or dechlorinating chemicals (included in the kit) can be added to tap water. In some water systems, chloramine, a newer additive, is used in place of chlorine, and it will not leave the water when exposed to air. You must use water conditioners that specifically say "removes chloramine." It might be beneficial to

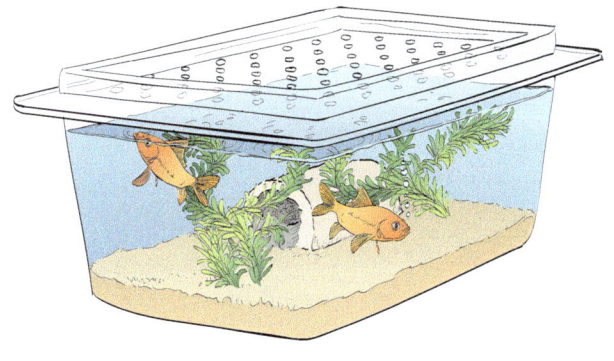

Animals Two by Two Module—FOSS Next Generation

INVESTIGATION 1 – *Goldfish and Guppies*

Goldfish

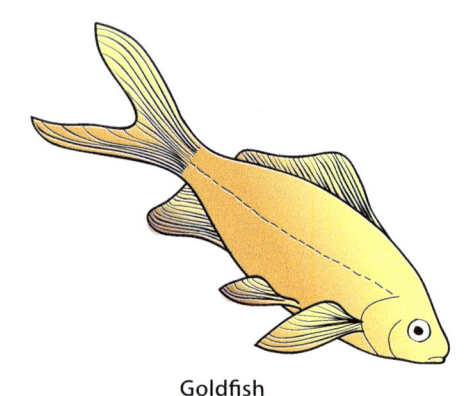

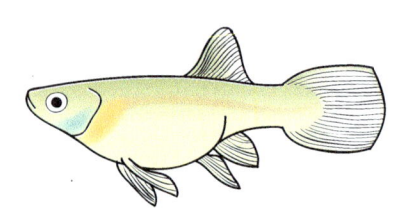

Female guppy

Male guppy

ask at your local pet store or aquarium store just what is recommended in your area to make the water safe for fish. Set aside a pitcher of aged or conditioned water so that you will have it ready to maintain the water level of the aquarium. Keep your aquarium covered to reduce evaporation and to keep dust out and fish in.

Adequate oxygen in the water is necessary to keep your fish healthy. Change about one-third of the aquarium water each week to maintain the chemical balance. If the water becomes cloudy or smelly, replace it immediately.

Feed the fish when students can observe the feeding behavior. Goldfish like to eat insect larvae, worms, aquatic plants, and snail eggs (all of which they eat in the wild) as well as commercial **food**. Guppies also eat commercial fish food, as well as finely chopped fish, tubifex worms, earthworms, and daphnia. A container of commercial fish food is provided in the kit. Feed your fish once or twice a day as much food as they will consume in 3–5 minutes. Too much food left in the aquarium will foul the water. Fish-feeding cakes—compressed food that disintegrates slowly—are available at pet and aquarium stores, if you need to leave the fish unattended for more than 3 days.

Goldfish and guppies generally do not require an aquarium heater, although you might want to purchase one if your classroom gets **below** 15°C on weekends. Fish cannot tolerate extremely warm water, either. Check the water temperature in hot weather. If the water is above 29.5°C, add cool water to lower the temperature.

The aquarium requires some light for the **plants** to grow but should not be in direct sunlight. Direct sunlight encourages algae growth, which turns the water green. Nutrients added to the water in the form of fish waste also encourage algae growth.

If a fish dies, you might want to leave it in the aquarium until students come to school, but dead fish foul the water, so they should be removed from the aquarium as soon as possible. Some students might be upset that a fish has died, and might need some comforting. Rather than simply throwing a dead fish away, take it outside and bury it in the ground near a plant. Discuss how this will help fertilize the plant, giving it additional nutrients to help it grow as the body of the fish decays.

Background for the Teacher

What Do Goldfish Do?

With 30,000 members in the fish clan, you might expect to find some extreme appearances and behavior. Some fish live part of their life in **fresh water** and part in salt water. This is an incredible feat, but trout, salmon, striped bass, shad, and others do it. **Flying** fish actually take to the air and glide on oversized fins to evade predators. Electric eels pack an electric jolt that can stun a person at a considerable distance, and other fish can produce light. Some fish can barely swim, **preferring** instead to walk about on the **bottom** of the sea on their fins. Pipefish, globefish, swordfish, sailfish, and flatfish are named for their particular physical appearance. The curious creatures called sea horses don't even look like fish, and the males brood the young in a pouch. There are fish so poisonous that one jab by a fin spine can cause death, and fish so grotesque that they could almost scare you to death. And when the most important life-sustaining part of a fish's environment threatens to dry up, some fish can burrow into the mud, wrap themselves in a secure envelope of mucus, and wait out the dry season in quiet repose. Another fish when faced with the **same** dire prospects sets out over land, looking for a place where the water stands a little deeper.

Adding a **tunnel** to the goldfish aquarium provides the opportunities for fish to swim **through** the tunnel, **behind** or **in front of** the tunnel—all relative positions that students can observe and describe.

> **NOTE**
> Live guppies and goldfish should not be released in ponds or creeks. Allow them to live out their short lives in your classroom.

INVESTIGATION 1 – *Goldfish and Guppies*

New Word — Say it, See it, Hear it, Write it

Above
Animal
Aquarium
Backward
Behind
Below
Bill
Bird
Bottom
Color
Compare
Different
Dirty
Eye
Feather
Female
Fin
Fish
Fly
Food
Forward
Fresh water
Gill
Goldfish
Guppy
Head
In front of
Male
Middle
Mouth
Next to
Plant
Prefer
Same
Scale
Structure
Surface
Swim
Tail
Through
Tunnel
Water
Wing

How Are Guppies and Goldfish Different? How Are They the Same?

Common goldfish are usually bright orange in **color**, but they can be white, black, or multicolored. The stock from which they are derived is a flat tan gray, and occasionally you will find a native-colored fish among the feeders. It is impossible to tell **males** and **females** apart, unless, of course, you are a goldfish.

Guppies are small fish that bear live young. The feeder-guppy females are larger and usually a uniform beige or silver-gray. Their abdomens become quite large when they are gravid (carrying young). The males are smaller and have longer, flowing tails. Males are the ones with spots of multiple colors. Fancy guppies that have been bred for showy colors can be dazzling.

Guppies are quite prolific and will probably give birth during their stay in your classroom. In fact, you might observe the arrival of baby guppies a day or two after the adults are put in their basin aquarium. The stress of transportation might induce a gravid female to release the babies. Adult guppies will eat the young, so you should supply the aquarium with plenty of elodea in which the babies can hide, or move the adults to a separate tank. Students will enjoy watching the baby guppies grow. Female goldfish lay eggs; they do not give live birth. Although they are prolific in nature, they usually will not breed in a small aquarium.

What Birds Visit Our Schoolyard?

Not every school has a pond with fish where students can go outdoors and experience animal life. Instead of looking for fish, students will go **bird** watching and find out what birds visit their schoolyard.

There are four key things to use for bird identification.

1. Size and overall shape of the bird. Three good birds to use as size references are a house sparrow (about 12.5 centimeters [cm]), an American robin (about 25 cm), and an American crow (43 cm).

2. Overall color pattern.

3. Behavior.

4. Habitat.

Full Option Science System

Background for the Teacher

Crow *Robin* *House sparrow*

Students might observe sparrows in bushes, robins feeding on the ground, and crows sitting on wires. What birds are doing and where they do it will be important things for students to consider. Students might also be able to see the size and shape of the **bill**, the markings on the **wings**, and the color of the **feathers**. These parts or structures are common to all birds.

Knowing the name of a specific bird is not as important as being able to describe what the bird generally looks like and what it does. But names do help you communicate to others and find out more about a specific kind of bird. A number of excellent online bird guides can provide information about common birds in different regions. Some of these resources provide videos and audio recordings so students can see and hear local birds in action. And there are probably a number of naturalists in your community who can provide lists of common birds and might even come and walk with your students as they watch birds in the schoolyard. Find those local naturalists and invite them to share their knowledge with your class.

Gull *Blue Jay* *Cardinal*

Animals Two by Two Module—FOSS Next Generation

INVESTIGATION 1 – *Goldfish and Guppies*

TEACHING CHILDREN about *Goldfish and Guppies*

Developing Disciplinary Core Ideas (DCI)

Just about every primary student knows something about fish. Most have seen fish in an aquarium at home or school, exotic fish on television, or perhaps large and numerous fish at a municipal aquarium. These experiences expand the early-childhood student's worldview, but the most valuable experience results from long-term observation and inquiry, where the student has responsibility for the well-being of the animal. Caring for other living things is a critically important character trait that can be developed with living organisms.

One of the keys to opening young minds to the natural world is the development of their powers of observation—not only the acuity of the sensory organs, but the disposition of mind for finding things out. Looking for fish in a tank and counting them is a good observation, but with guidance young students will observe subtlety of shape and color and nuance of behavior. Once students are tuned into good observations, they will be able to make the same kinds of observations in other settings and compare the observations. Through comparisons, students make discoveries, and discoveries are memorable.

Primary science is a delicate balance between instruction and setting limits. The natural inclination of kindergartners is to wade into new situations and start testing. We don't want to stymie this enthusiasm for discovery, but at the same time we want to direct young students into socially productive behavior. There are things we want them to know about the world and how it is organized. We want to tell them, show them, teach them. But science isn't well received in early childhood if it is structured your way—it should be structured their way. Provide new materials and let students explore. The structure is more in the form of setting limits than of providing information. Where living organisms are concerned, students might want to try some "experiments" that are inappropriate and harmful to the organism. Setting strict limits is important.

A good way to draw young students into a subject is by asking questions, rather than issuing instructions. The student inquiry into fish might start with "Boys and girls, look what we have for our classroom. Fish! What are they doing? Where do you think we should put the aquarium?" As interest wanes, a new item in the aquarium and new questions can rekindle interest. Representational materials (books) provide excellent enrichment for young students *after* they have had firsthand experiences.

NGSS Foundation Box for DCI

LS1.A: Structure and function
- All organisms have external parts. Different animals use their body parts in different ways to see, hear, grasp objects, protect themselves, move from place to place, and seek, find, and take in food, water, and air. Plants also have different parts (roots, stems, leaves, flowers, fruits) that help them survive and grow. (foundational)

LS1.C: Organization for matter and energy flow in organisms
- All animals need food in order to grow. They obtain their food from plants or from other animals. Plants need water and light to live and grow. (K-LS1-1)

ESS2.E: Biogeology
- Plants and animals can change their environment. (K-ESS2-2)

ESS3.A: Natural resources
- Living things need water, air, and resources from the land, and they live in places that have the things they need. Humans use natural resources for everything they do. (K-ESS3-1)

Teaching Children about Goldfish and Guppies

The activities and readings students experience in this investigation contribute to disciplinary core ideas **LS1.A, Structure and function:** All organisms have external parts; **LS1.C, Organization for matter and energy flow in organisms:** All animals need food in order to grow; **ESS2.E, Biogeology:** Plants and animals can change their environment; and **ESS3.A, Natural resources:** Living things need water, air, and resources from the land and they live in places that have the things they need.

Engaging in Science and Engineering Practices (SEP)

Engaging in the practices of science helps students understand how scientific knowledge develops; such direct involvement gives them an appreciation of the wide range of approaches that are used to investigate, model, and explain the world. Engaging in the practices of engineering likewise helps students understand the work of engineers, as well as the links between engineering and science. Participation in these practices also helps students form an understanding of the crosscutting concepts and disciplinary ideas of science and engineering; moreover, it makes students' knowledge more meaningful and embeds it more deeply into their worldview. (National Research Council, *A Framework for K–12 Science Education*, 2012, page 42)

In this investigation, students engage in these practices.

- **Asking questions** about different kinds of fish and birds and the places where they live and get resources.
- **Developing and using a model** of a fish aquarium to show the relationship between the animals and their environment.
- **Planning and carrying out investigations** in collaboration with peers and with adult guidance involving fish and birds to observe their structures and study their environmental needs; make predictions based on prior experiences.
- **Analyzing and interpreting data** by describing observations of the fish over time, recording information, using and sharing notebook entries, including writing and labeled pictures. Students use their firsthand observations and those of others in the classroom to describe the patterns they observe in fish aquaria to answer scientific questions. Students compare their predictions to actual outcomes.
- **Constructing explanations** by making firsthand observations of fish and birds and using this as evidence to answer questions about the needs of animals, including food; supporting arguments with evidence.

NGSS Foundation Box for SEP

- **Ask questions** based on observations to find more information about the natural and/or designed world(s).
- **Use a model** to represent relationships in the natural world.
- **With guidance, plan and conduct an investigation** in collaboration with peers (for Grade K).
- **Make observations** (firsthand or from media) and/or measurements to collect data that can be used to make comparisons.
- **Make predictions** based on prior experiences.
- **Record information** (observations, thoughts, and ideas).
- **Use and share pictures, drawings,** and/or writings of observations.
- **Use observations (firsthand or from media)** to describe patterns in the natural world in order to answer scientific questions.
- **Compare predictions** (based on prior experiences) to what occurred (observable events).
- **Make observations (firsthand or from media)** to construct an evidence-based account for natural phenomena.
- **Construct an argument** with evidence to support a claim.
- **Read grade-appropriate text** and/or use media to obtain scientific and/or technical information to describe patterns in the natural world.
- **Communicate** information or solutions with others in oral and/or written forms using models and/or drawings that provide detail about scientific ideas.

INVESTIGATION 1 – *Goldfish and Guppies*

> **NGSS Foundation Box for CC**
>
> - **Patterns:** Patterns in the natural and human-designed world can be observed, used to describe phenomena, and used as evidence.
> - **Cause and effect:** Events have causes that generate observable patterns. Simple text can be designed to gather evidence to support or refute student ideas about causes.
> - **Systems and system models:** Objects and organisms can be described in terms of their parts. Systems in the natural and designed world have parts that work together.
> - **Structure and function:** The shape and stability of structures of natural and designed objects are related to their function(s).

- **Obtaining, evaluating, and communicating information** about structures of fish and birds, their needs, and where they live by reading grade-appropriate text and communicating in oral and written formats.

Exposing Crosscutting Concepts (CC)

The third dimension of instruction involves the crosscutting concepts, sometimes referred to as unifying principles, themes, or big ideas, that are fundamental to the understanding of science and engineering.

These concepts should become common and familiar touchstones across the disciplines and grade levels. Explicit reference to the concepts, as well as their emergence in multiple disciplinary contexts, can help students develop a cumulative, coherent, and usable understanding of science and engineering. (National Research Council, 2012, page 83)

In this investigation, the focus is on these crosscutting concepts.

- **Patterns.** Structures of fish are similar but they have differences in how they look and where they live.
- **Cause and effect.** Fish can change their water environment over time.
- **Systems and system models.** Fish and birds can be described in terms of their structures.
- **Structure and function.** The observable structures of fish (head, eyes, mouth, fins, tail) and birds (head, eyes, beak, wings, tail) serve functions in survival.

Connections to the Nature of Science

This investigation provides connections to the nature of science.

- **Scientific investigations use a variety of methods.** Scientific investigations begin with a question. Scientists use different ways to study the world.
- **Scientific knowledge is based on empirical evidence.** Scientists look for patterns and order when making observations about the natural world.

Teaching Children about Goldfish and Guppies

Conceptual Flow

The anchor phenomenon for this module is that animals of all kinds have various needs to live and grow—including humans. To understand an animal's needs, you need to first get to know the animal—its structures and behaviors. The first animals students study and care for live in water. Fish are the main phenomenon in this experience. The guiding question is what do animals such as fish and birds need to live and grow?

A story about a fish that has to swim a great distance from the sea to a river is more meaningful if students know how fish swim. Swimming up a river is not an abstraction for students; it is the sinuous action that they have observed firsthand, accompanied by the movements of a variety of fins. Similarly, video adventures with fish from around the world mean more when students have a concept of "fishness" developed through personal experience. In FOSS, our recommendation is that representational materials be used after firsthand experiences.

Although the *Investigations Guide* and center instructions suggest ways to guide student observations, you might find that students are so excited and curious about the animal visitors that you won't get a chance to ask the questions suggested. We recommend that you follow the lead of the students. The structure of the lesson should always remain flexible, especially with living organisms. You never know exactly what they're going to do!

The **conceptual flow** for this first investigation starts with an introduction to a goldfish. Students observe several goldfish, describe their parts (structures), and compare the variations in size and color. In Part 2, students learn how to care for fish, providing them with food and fresh water, and enrich the aquarium with plants.

In Part 3, students add a tunnel to the aquarium to observe how the fish respond. In Part 4, students study a new kind of fish and compare the guppy **structures and behaviors** to the goldfish.

In Part 5, students continue their study of animal structures, behaviors, and needs by going to the schoolyard to observe common birds. They make several trips to observe what the birds look like, what they are doing, and what they need to live.

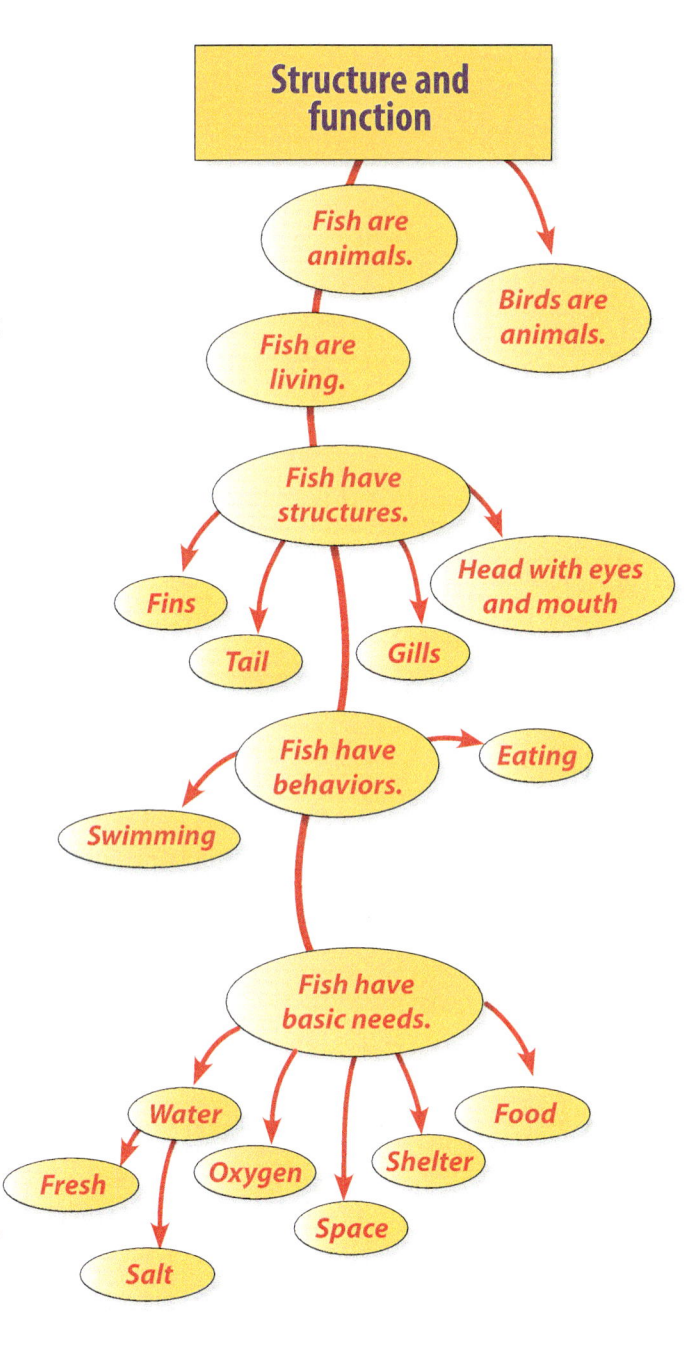

Animals Two by Two Module—FOSS Next Generation

INVESTIGATION 1 – *Goldfish and Guppies*

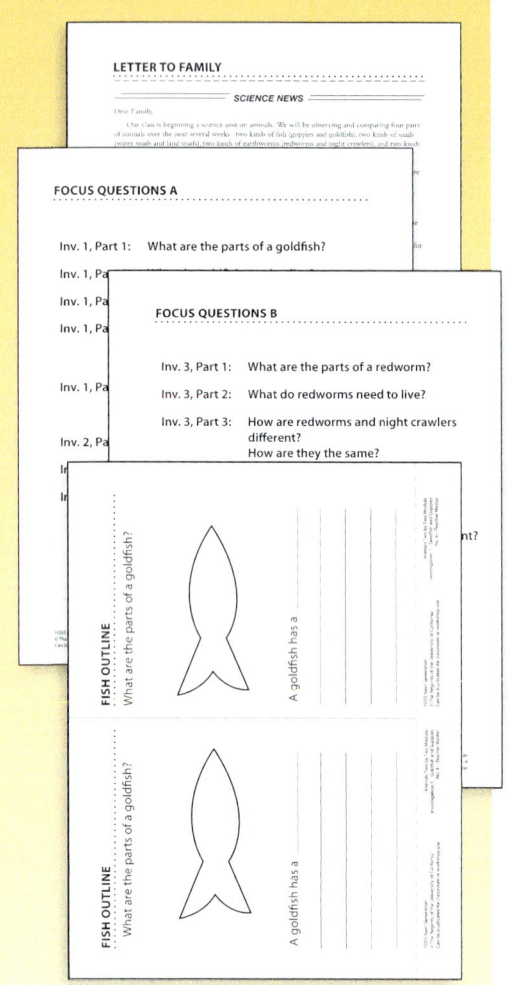

Nos. 1–4—Teacher Master

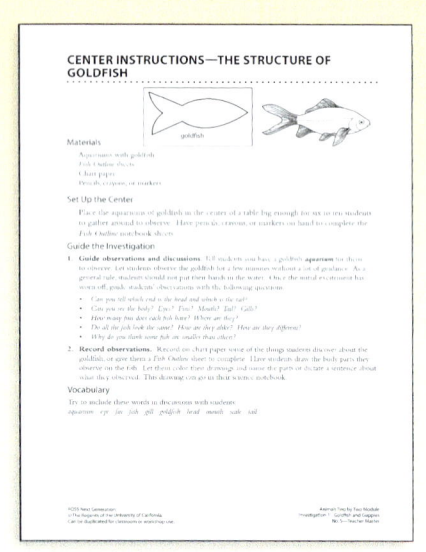

No. 5—Teacher Master

MATERIALS for
Part 1: *The Structure of Goldfish*

For each student
- 1 *Letter to Family* ★
- 1 Science notebook (See Step 9 of Getting Ready.) ★
- • Crayons and pencils ★
- 1 *Fish Outline* ★

For the class
- • Chart paper ★
- 3 Clear basins with covers
- 1 Collecting net
- 1 Vial of fish food
- 1 Bottle of dechlorinator
- 2 Self-stick notes
- • Pencils, crayons, or markers ★
- 2 Goldfish (See Step 5 of Getting Ready.) ★
- 6 Guppies (See Step 5 of Getting Ready.) ★
- 1 Bunch of elodea, 6–8 sprigs ★
- • Aged water (See Step 3 of Getting Ready.) ★
- 1 Gravel, bag, 1 kg/bag
- 1 Pitcher or water container ★
- • *FOSS Science Safety*, *FOSS Outdoor Safety*, and *Conservation* posters
- ❏ 1 Teacher master 1, *Letter to Family*
- ❏ 1 Teacher master 2–3, *Focus Questions A and B*
- ❏ 1 Teacher master 4, *Fish Outline*
- ❏ 1 Teacher master 5, *Center Instructions—The Structure of Goldfish*

For assessment
- ❏ • *Assessment Checklists* 1 and 2

★ Supplied by the teacher.   ❏ Use the duplication master to make copies.

78   Full Option Science System

Part 1: The Structure of Goldfish

GETTING READY for
Part 1: *The Structure of Goldfish*

1. **Schedule the investigation**
   This part requires 15–20 minutes at the center for each group of six to ten students. Plan 5 minutes to introduce and 5 minutes to wrap up the session with the entire class at the rug. Plan 15 minutes for students to write or draw in their notebooks.

2. **Preview Part 1**
   Students observe goldfish living in a simple aquarium. They look for and name different parts of the fish, such as fins, tail, mouth, and gills. They look to see if all the fish are alike, or if there are differences such as color and size. They draw a picture and dictate a sentence to record what they see. The focus question is **What are the parts of a goldfish?**

3. **Prepare water for the aquariums**
   Age tap water for 24 hours in an open container, or use the chemical dechlorinator to prepare the water. You will need approximately 5 L of water for each of three aquariums (two for goldfish and one for guppies). Keep an additional bottle of conditioned water handy for replenishing the basins.

4. **Set up the aquariums**
   You can set up all the aquariums at the same time, but keep the guppies out of sight until Part 4. Rinse the bag of gravel well, removing the fine dust/sand particles. Place 1/4 of the gravel into each aquarium. Fill the containers with about 5 L of conditioned water. Mark the water level on each container with a self-stick note to keep track of how much has evaporated and needs replacing.

5. **Obtain fish and elodea**
   Look online for "Aquariums and Supplies," "Pet Supplies," or "Tropical Fish" to find a local store where you can obtain goldfish and guppies. Ask for common, nonfancy, healthy, inexpensive goldfish and guppies. Although feeder goldfish are less expensive, it is our experience that they do not do as well in the classroom. Ask the dealer to try to give you four female and two male guppies. They are easily identified, as noted in the background information. Goldfish gender, however, cannot be identified by their appearance.

   These stores also sell an aquatic plant called elodea *(Anacharis)* by the bunch. You'll need at least two sprigs for each aquarium setup. If elodea is not available, use another rooted aquatic plant.

TEACHING NOTE

The Getting Ready section for Part 1 of Investigation 1 is longer than the corresponding section for the other parts. Several of the numbered items appear only here but may apply to other parts as well. These include planning a word wall (Step 8), printing or photocopying duplication masters, (Steps 9–10), and planning for safety (Step 12).

▶ **NOTE**
To prepare for this investigation, view the teacher preparation video on FOSSweb.

▶ **NOTE**
Some areas have chloramine in their tap water. In this case, the dechlorinator must be used to treat the water.

▶ **NOTE**
One-fourth of the gravel will be used later for the water snail aquarium.

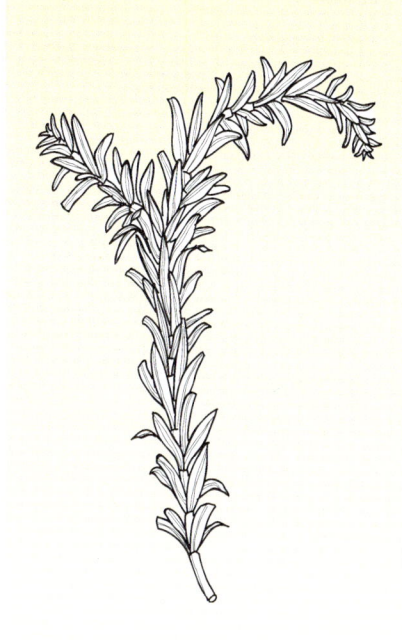

Animals Two by Two Module—FOSS Next Generation

INVESTIGATION 1 – Goldfish and Guppies

TEACHING NOTE

See Preparing the Kit for Your Classroom in the Materials chapter for more information.

6. **Introduce the fish into the aquariums**
   The fish will probably be in plastic bags for transportation. Float the bags, fish and all, in the aquariums for about 15 minutes to equalize the temperatures. Then release goldfish in two aquariums, and guppies in the third (goldfish will eat guppies). For the first session, put all the elodea in with the guppies.

   Keep the guppies out of sight until later in the week. Feed and care for them each day.

7. **Set up the center**
   The fish should be set up at a center that small groups of students can visit at various times during the day. Place the aquariums of goldfish in the center of a table big enough for six to ten students to gather around to observe. (The aquariums should initially contain only water and goldfish, but it will be enriched as the investigation progresses.) You might eventually want to purchase a fishbowl or aquarium with a filter and have fish become a permanent part of your classroom.

8. **Plan to use a word wall or pocket chart**
   As the module progresses, you will add new vocabulary words to cards or sentence strips for use in a pocket chart and/or to a word-wall chart for posting on a wall or an easel. You will also use a chart for writing the day's focus question and for modeling responses. For additional information, see the Science-Centered Language Development chapter in *Teacher Resources*.

FOCUS CHART

What are the parts of a goldfish?

Goldfish have two eyes, one tail, six fins, one mouth, gills, and scales.

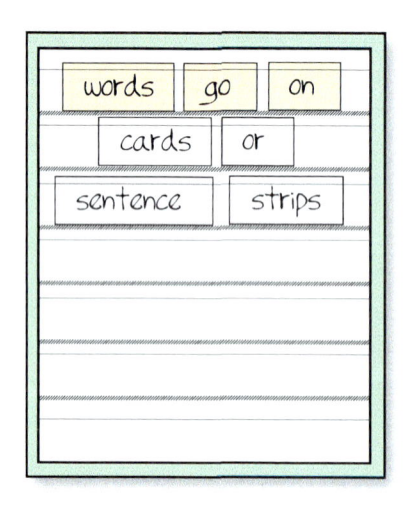

---

words go on
cards or
sentence strips

WORD WALL

words
words
and more words

---

80 Full Option Science System

Part 1: The Structure of Goldfish

9. **Plan for student notebooks**

   Each student will keep a record of science investigations in his or her science notebook. Students will record observations and responses to focus questions. This record will be a useful reference document for students and a revealing testament for adults of each student's learning progress.

   We recommend that students use bound composition books for their science notebooks. This ensures that student work is maintained as an organized, sequential record of student learning.

   You can make individual science notebooks by stapling 12–15 blank pages together with a cover. If you are already using an alternative method of organization with your students, such as a sheaf of folded and stapled pages, your method can take the place of the bound composition book.

   Each part of each investigation starts with a focus question. You'll find these questions on teacher masters 2 and 3, *Focus Questions A* and *B*. There are a couple of ways to provide these questions to students.

   - Print or photocopy the teacher masters, cut the questions apart, and have them ready for each student to glue into his or her notebook at the appropriate time in each part.
   - Prepare sheets with one focus question on each sheet and room for students to draw and write. You could also include writing frames on each sheet. Print or photocopy these sheets and assemble them into a science notebook for each student. You can print the focus questions on address labels for students to place in their notebooks.

   TEACHING NOTE

   See the Science Notebooks in Grades K–2 chapter for details on setting up and using notebooks.

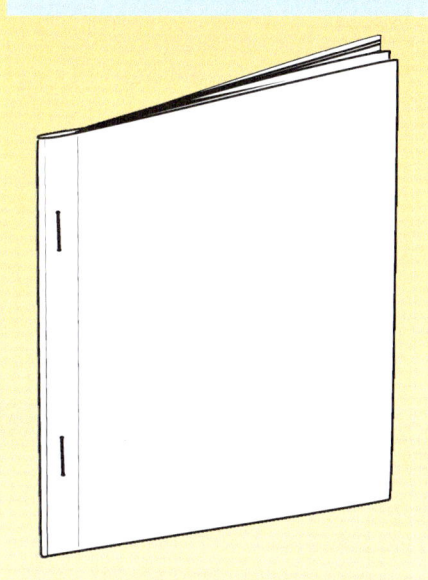

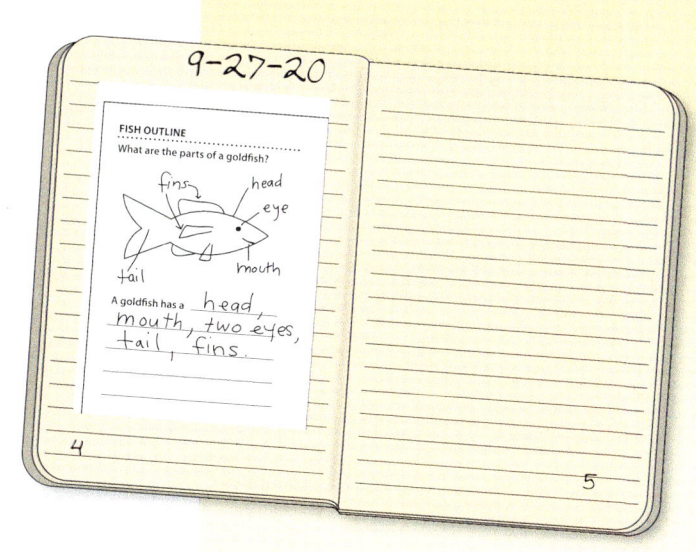

10. **Print or photocopy duplication masters**

    Teacher masters serve various functions—letter to family, home/school connections, center instructions, notebook sheets, and focus questions. These can be duplicated, and some will be cut apart and glued into student notebooks. A master that requires printing or duplication is flagged with this icon ❏ in the materials list for each part.

11. **Send a letter home to families**

    Read teacher master 1, *Letter to Family*, and send it home with students as you begin this investigation. The letter explains what students will be doing with animals and what parents can do to extend the experiences at home with their children.

Animals Two by Two Module—FOSS Next Generation

81

INVESTIGATION 1 – *Goldfish and Guppies*

TEACHING NOTE

For a detailed discussion of strategies for working outdoors with students, see the Taking FOSS Outdoors chapter.

EL NOTE

Project or post the equipment photo card for each object, and write the object's name on the word wall. Equipment photo cards are available on FOSSweb and can be downloaded and printed.

12. Plan for safety indoors and outdoors

Young students must be allowed to demonstrate that they can act responsibly with materials, but they must be given guidelines for safe and appropriate use of materials. Work with students to develop those guidelines so that students participate in making behavior rules and understand the rationale for the rules. Encourage responsible actions toward other students. Display and discuss the *Science Safety* and *Outdoor Safety* posters in class.

Look for the safety icon in the Getting Ready and Guiding the Investigation sections, which alerts you to safety concerns throughout the module. Be aware of any allergies your students might have.

This is a good time to introduce the set of four *Conservation* posters and discuss the importance of natural resources with students.

13. Plan for working with English learners

At important junctures in an investigation, you'll see sidebar notes titled "EL Note." These notes suggest additional strategies for enhancing access to the science concepts for English learners. These strategies are helpful for all kindergarten students. Refer to the Science-Centered Language Development chapter for resources and examples to use when working with science vocabulary, writing, oral discourse, and readings.

Each time new science vocabulary is introduced, you'll see the new-word icon in the sidebar. This icon lets you know not only that you'll be introducing important vocabulary, but also that you might want to plan on spending more time with those students who need extra help with the vocabulary.

14. Assess progress throughout the module

Assessment opportunities are embedded throughout the module to help you look closely at students' progress. In kindergarten, those assessment opportunities involve teacher observations or performance observations of students' actions with materials as well as teacher-student and student-student verbal interactions, focusing on the ideas under study. Read through the Assessment chapter for a description of the assessment opportunities.

You will find the *Assessment Checklists* in the Assessment Masters chapter in *Teacher Resources*. *Assessment Checklist* no. 1 lists the disciplinary core ideas that students will encounter throughout

Part 1: The Structure of Goldfish

the module and that are described in each Getting Ready section. You can assess each of these objectives several times during the course of the module. *Assessment Checklist* no. 2 is for recording students' ability to engage in science and engineering practices appropriate for their age. *Assessment Checklist* no. 3 is for recording opportunities to engage with crosscutting concepts and during each investigation. Plan to use this chart at the end of each investigation.

15. Plan assessment for Part 1

Print or make copies of these assessment checklists to use throughout this investigation. Most of these checklist copies can be used for other investigations as well.

- *Assessment Checklist* 1: Disciplinary Core Ideas
- *Assessment Checklist* 2: Science and Engineering Practices
- *Assessment Checklist* 3: Crosscutting Concepts

What to Look For

- *Students handle living things with respect as they make and record observations of the structures of fish. (Planning and carrying out investigations; structure and function.)*

- *Students begin to construct the core idea that animals (fish) have external parts that help them meet their needs. (LS1.A: Structure and function.)*

Attach the checklists to a clipboard and carry them with you when students are engaged in the investigation. Record your observations on the appropriate assessment checklist as you interact with students, or take a few minutes after class to reflect on the lesson..

During each part of each investigation, FOSS suggests focusing on one disciplinary core idea, one science and engineering practice, and one crosscutting concept, but you are not restricted to these if you see something else worth noting. All checklists contain all the possibilities for the three dimensions of science learning.

Plan to listen and observe students as they work and to review their notebook entries during or after class.

No. 1—Assessment Master

Animals Two by Two Module—FOSS Next Generation

INVESTIGATION 1 – Goldfish and Guppies

FOCUS QUESTION
What are the parts of a goldfish?

Materials for Step 1
- *Goldfish aquariums*

SCIENCE AND ENGINEERING PRACTICES

Asking questions

Planning and carrying out investigations

Analyzing and interpreting data

TEACHING NOTE

The gills of a fish are located inside the body. Students can observe the gill covers or slits that protect the gills.

animal
aquarium
eye
fin
gill
goldfish
head
mouth
scale
tail
water

Materials for Step 4
- ***Fish Outline** sheets*
- *Crayons, pencils, or markers*

GUIDING the Investigation
Part 1: The Structure of Goldfish

1. **Introduce the investigation**
   Call the class to the rug. Show students one **aquarium** container with **water** and **goldfish**. Ask them what they think is in the container.

   If students don't know what kind of fish they are, tell them that these are goldfish. Goldfish are **animals**. Tell students,

   Over the next few weeks, we will be answering questions about fish—what they look like, what they do, and what they need to live. The goldfish will be at the learning center. When it is your turn today, you should observe the fish to see what they look like. For example, what is the shape of their body, and what other parts can you see?

   Choose groups to rotate through the center, or adjourn to free-choice time.

2. **Guide observations and discussions**
   At the center, let students observe the goldfish for a few minutes without a lot of guidance. As a general rule, they should not put their hands in the water. Once the initial excitement has worn off, guide students to make observations. Start by having them tell you what they notice. Use these questions as a guide.

   ➤ *Can you tell which end is the **head** and which is the **tail**?*

   ➤ *Can you see the body? **Eyes**? **Fins**? **Mouth**? Tail? **Gills**? **Scales**?*

   ➤ *How many fins does each fish have? Where are they?*

   ➤ *Do all the fish look the same? How are they alike? How are they different?*

   ➤ *Why do you think some fish are smaller than others?*

   ➤ *What questions do you have about goldfish?*

3. **Review vocabulary**
   As students offer their observations, add any new or important vocabulary to the class word wall. Let students be the guides—acknowledge the words they use and offer new vocabulary as needed.

4. **Focus question: What are the parts of a goldfish?**
   Write the focus question on the chart as you read it aloud.

   ➤ *What are the parts of a goldfish?*

   Tell students that you have a sheet with the question written on it. Give each student a copy of teacher master 4, *Fish Outline,* to draw

84   Full Option Science System

Part 1: The Structure of Goldfish

the goldfish structures they observe, such as eyes, fins, and tail, or have students make a drawing of the goldfish on their own. Model how to draw by looking at the different shapes that make up the goldfish body structures.

Have students dictate a sentence for the bottom of the sheet or direct them to choose words to add from the class word wall. Describe and model how to glue the sheet into their notebooks when they are finished.

5. **Model responses to the focus question (optional)**
Depending on students' experiences with notebooks, you can let them work on their own, or you can model making a notebook entry, using the focus chart.

WRAP-UP/WARM-UP

6. **Share notebook entries**
Conclude Part 1 or start Part 2 by having students share notebook entries. Ask students to open their science notebooks to the first entry. Read the focus question together.

➤ *What are the parts of a goldfish?*

Ask students to pair up with a partner to
- share their answers to the focus question;
- explain their drawings.

Ask students to think about and discuss what they think the goldfish parts are used for. [Eyes for seeing; fins and tail for swimming; mouth for eating; gills for breathing; scales for protection.] Some students will need more observations to connect parts (structures) with their function. Revisit that in the next part.

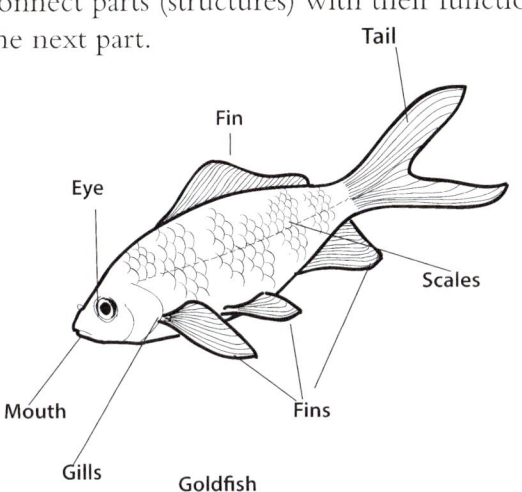

CROSSCUTTING CONCEPTS

Systems and system models
Structure and function

FOCUS CHART

What are the parts of a goldfish?

Goldfish have two eyes, one tail, six fins, one mouth, gills, and scales.

TEACHING NOTE

The focus question in each part engages students with the phenomenon to investigate.

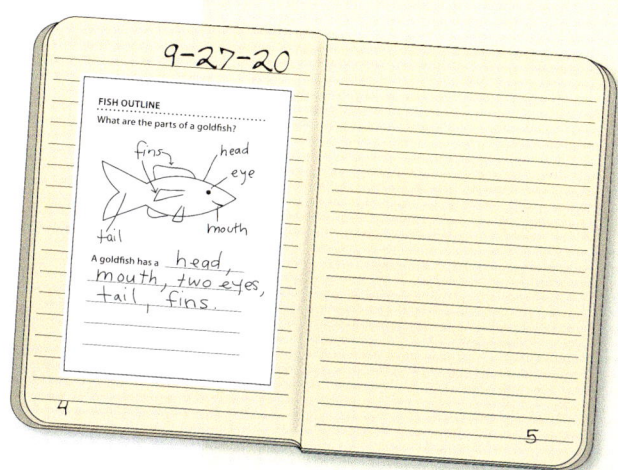

Animals Two by Two Module—FOSS Next Generation

INVESTIGATION 1 – *Goldfish and Guppies*

MATERIALS
Part 2: *Caring for Goldfish*

For the class

- 2 Goldfish aquariums (from Part 1)
- 1 Vial of fish food
- 1 Piece of scratch paper ★
- 1 Bunch of elodea, 6–8 sprigs ★
- 1 Plastic cup
- 1 Clear basin
- • Paper towels (optional) ★
- • Aged water ★
- 1 Collecting net
- ❏ 1 Teacher master 2, *Focus Questions A*
- ❏ 1 Teacher master 6, *Center Instructions—Caring for Goldfish*

For assessment

- • *Assessment Checklists* 1 and 2

★ Supplied by the teacher.   ❏ Use the duplication master to make copies.

No. 6—Teacher Master

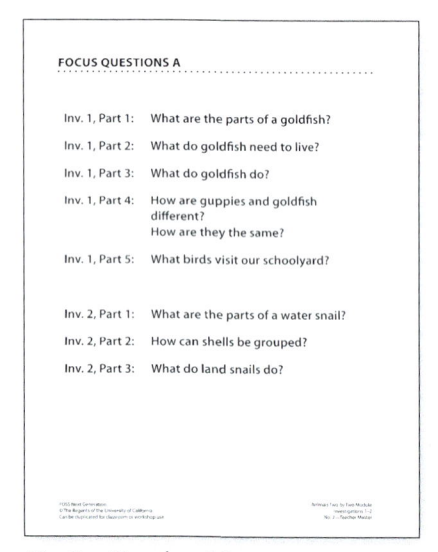

No. 2—Teacher Master

86 Full Option Science System

Part 2: Caring for Goldfish

GETTING READY *for*
Part 2: *Caring for Goldfish*

1. **Schedule the investigation**
   This part requires 5 minutes for the introduction and 15–20 minutes at the center for each group of six to ten students. If possible, each group should do this part on a different day, so that the fish will be hungry. It works well to alternate between Parts 2 and 3 on the same day with different groups. Plan 15 minutes for students to write or draw in their notebooks.

2. **Preview Part 2**
   Students learn how to care for goldfish, giving them food and fresh water, and adding plants to the aquarium. With each addition, students describe the fish behavior they observe. The focus question is **What do goldfish need to live?**

3. **Don't feed the fish**
   Don't feed the fish just prior to conducting this part. Students will be looking at feeding behavior, so you want the fish to be hungry.

4. **Have elodea handy**
   Students will add the aquatic plant elodea to the aquariums. The plants will go in and out of the aquarium several times, so that each group will have the experience of adding the plants. Be sure to keep the plants close at hand in a container of water or wrapped in wet paper towels until they are put into the aquarium with the goldfish.

5. **Have conditioned water handy**
   Students will remove two cups of water and replace it with new water that has been aged or conditioned. Have an empty basin and cup handy for removing water. Prepare new water by allowing it to age overnight or by using conditioner.

6. **Plan assessment for Part 2**
   Plan to observe students as they provide for the needs of fish.

   What to Look For

   - Students know that fish are animals that have basic needs—food to live and grow, water with oxygen, and space. (LS1.C: Organization for matter and energy flow in organisms; ESS2.E: Biogeology.)
   - Students ask questions about fish behavior based on their firsthand observations. (Asking questions.)
   - Students predict what changes will occur in the aquarium water. (Cause and effect.)

Animals Two by Two Module—FOSS Next Generation

INVESTIGATION 1 – Goldfish and Guppies

FOCUS QUESTION
What do goldfish need to live?

Materials for Step 2
- *Goldfish aquariums*

Materials for Steps 3–4
- *Fish food*
- *Scratch paper*
- *Plastic cup*
- *Empty basin*
- *Conditioned water*
- *Collecting net*

▶ **NOTE**
Feed the fish a tiny pinch of food twice a day. One feeding can be by the students, and the other by the teacher after class.

▶ **NOTE**
Use this opportunity to scoop out with the collecting net any excessive amounts of food students have added.

CROSSCUTTING CONCEPTS
Cause and effect

GUIDING the Investigation
Part 2: Caring for Goldfish

1. **Focus question: What do goldfish need to live?**
   Call the class to the rug. Write the focus question on the chart as you read it aloud.

   ➤ *What do goldfish need to live?*

   Give students time to share their ideas with a partner. Then tell them that over the next few days, they will find out what goldfish do when **food**, **fresh water**, and **plants** are added to their home.

2. **Call one group to the center**
   Let six to ten students spend some time watching the fish with adult supervision but without adult guidance. See if they notice any new parts they didn't see before.

3. **Feed the fish**
   Tell students,

   Goldfish need only a tiny bit of food every day. We must be careful not to put too much food in the water or it might make the goldfish sick.

   Put a small pinch of fish food on a piece of paper near the aquariums. Let each student take a tiny bit of the food and put it into one of the aquariums. Tell students to watch to see what the fish do. Ask them to think about and share with the group what they notice and what questions they have. If students need prompting, ask,

   ➤ *How do goldfish find their food? Do you think they smell it, see it, or just happen to run into it?*

   ➤ *Where do they eat their food? Are they top (**surface**), **middle**, or **bottom** feeders?*

   ➤ *Do fish chew their food?*

   ➤ *Do they eat all of their food? What happens to the food that they don't eat?*

   ➤ *What kinds of food do you think goldfish find to eat when they live in the wild?* [Insect larvae, worms, plants, snail eggs.]

4. **Renew the water**
   Ask,

   ➤ *How do we know when the water is **dirty**? Do you see anything in the water? At the bottom of the aquarium?*

   ➤ *How do you think it becomes dirty?*

88   Full Option Science System

Part 2: Caring for Goldfish

▶ *What should we do about it?*

Explain that goldfish need water that has oxygen for them to breathe. If the water is dirty, it may be a sign that it is low in oxygen. Students can scoop out some of the old water and replace it with clean fresh water that has been conditioned to remove chemicals that are not good for fish. Water that is good for people to drink may need to be treated to make it suitable for fish to live in. Have students help remove two cups of the aquarium water, dumping it into a basin.

Before adding the fresh water, give students a few minutes to make predictions about what will happen when the clean water is added. Students can turn and talk with a neighbor using this sentence frame, I think ____ because ____ .

Then add new, conditioned water.

▶ *What do the **fish** do when the water is added?*

▶ *Do they go toward the new water or away from it?*

5. Add plants to the aquarium

Have students add several sprigs of elodea to one end of each aquarium. Encourage students to ask questions about goldfish as they observe and describe how the fish respond to the plants. Ask,

▶ *Do they touch the plants? What part of their bodies do they use to find out about the plants?*

▶ *Do the goldfish **prefer** to **swim** near the plants, or do they prefer the open spaces? How do you know?*

▶ *Do they swim around the plants, **through** the plants, or **behind** the plants?*

▶ *If this was a natural pond outdoors, and there were bigger fish and birds who might eat the goldfish, where do you think the goldfish would find shelter and hide?*

6. Guide additional observations

After a few minutes, ask students to observe how fish move.

▶ *What parts of the goldfish's body help it move around?*

▶ *How many different directions can it move? **Forward**? **Backward**? Sideways? Up? Down?*

▶ *Do goldfish ever rest, or are they always moving?*

▶ *What questions do you have about goldfish?*

Materials for Step 5
- *Elodea*

SCIENCE AND ENGINEERING PRACTICES

Asking questions

Analyzing and interpreting data

CROSSCUTTING CONCEPTS

Structure and function

Animals Two by Two Module—FOSS Next Generation

INVESTIGATION 1 – *Goldfish and Guppies*

backward
behind
bottom
dirty
fish
food
forward
fresh water
middle
plant
prefer
surface
swim
through

7. Review vocabulary

As students offer their observations, add any new or important vocabulary to the class word wall. Acknowledge the words students use and offer new vocabulary as needed.

Have students recall what fish need to live. Before leaving the center, ask each student to tell you one thing he or she observed.

8. Have a sense-making discussion

Sense-making discussions can be conducted in small groups as part of center time or as a whole class once all students have completed the center activity. The intent is for students to have a sense-making discussion after working with materials and before answering the focus question.

➤ *What do goldfish need to live?*

➤ *What if we put too much food in the water?*

➤ *What would happen if there wasn't enough food or water?*

9. Answer the focus question

Restate the focus question, and have the class read it aloud together.

➤ *What do goldfish need to live?*

Tell students that you have a strip of paper with the focus question written on it. Describe how to glue the strip into the notebook before they answer the question. Ask them to return to their tables and work in their notebooks. For emerging writers you can provide a sentence frame such as, Fish need _____ to live.

Students can also draw pictures to show what goldfish need to live.

10. Prepare for the next group

Remove the plants before the next group arrives.

WRAP-UP/WARM-UP

11. Share notebook entries

Conclude Part 2 or start Part 3 by having students share notebook entries. Read the focus question together.

➤ *What do goldfish need to live?*

Ask students to pair up with a partner to

- share their answers to the focus question;
- explain their drawings.

Ask students to think about and discuss how the classroom goldfish gets what it needs compared to how a goldfish outside in a pond gets food, clean water, and space with shelter.

FOCUS CHART

What do goldfish need to live?

Goldfish need food, clean water, and space with shelter.

Part 3: Goldfish Behavior

MATERIALS for
Part 3: Goldfish Behavior

For each student
1 Envelope (optional) ★
1 *Paper Aquarium*

For the class
2 Goldfish aquariums with plants
2 Fish tunnels, square plastic pipe
• Transparent tape ★
1 Paper cutter (optional) ★
10 Scissors ★
❏ 1 Teacher master 7, *Center Instructions—Goldfish Behavior*
❏ 1 Teacher master 8, *Paper Aquarium*

For assessment
• *Assessment Checklists* 1 and 2

★ Supplied by the teacher.   ❏ Use the duplication master to make copies.

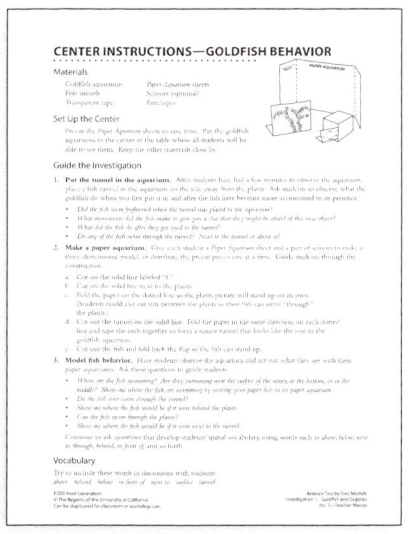

No. 7—Teacher Master

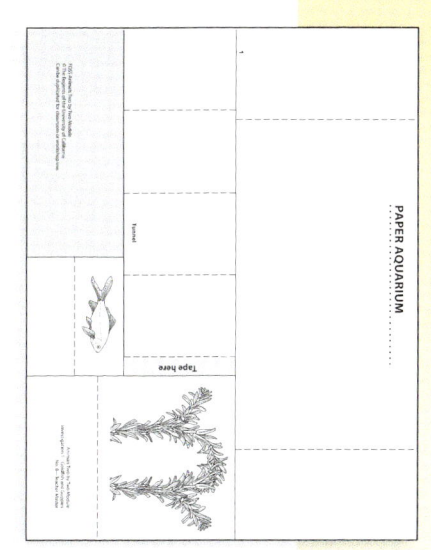

No. 8—Teacher Master

Animals Two by Two Module—FOSS Next Generation

91

INVESTIGATION 1 – *Goldfish and Guppies*

GETTING READY for
Part 3: *Goldfish Behavior*

1. **Schedule the investigation**
   Each group of six to ten students will need 10–15 minutes at the center to observe the fish with the tunnel. Plan 5 minutes for introduction and 15 minutes for students to write or draw in their notebooks.

2. **Preview Part 3**
   Students add a tunnel to the aquarium to observe how the fish respond. They make their own paper aquariums to model the fish behavior they have observed. The focus question is **What do goldfish do?**

3. **Cut the paper aquariums (optional)**
   If center time is limited, use a paper cutter to precut the paper aquarium sheets into four parts. Students will have more time to create the props and model fish behavior.

4. **Plan assessment for Part 3**
   Plan to observe students as they modify the fish aquarium and model what happens.

   What to Look For

   - *Students use a model of a fish aquarium to accurately show the relationship between the fish and the place where it lives. (Developing and using models; systems and system models.)*

   - *Students describe observations of fish behavior in the aquarium when a tunnel is added. (Cause and effect.)*

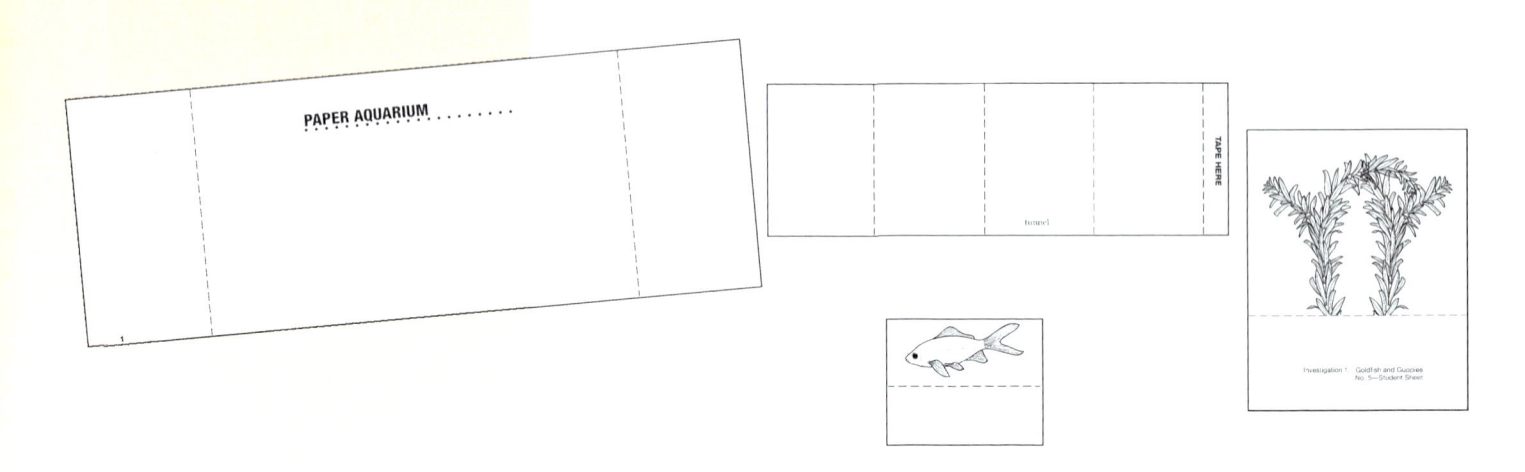

92 Full Option Science System

Part 3: Goldfish Behavior

GUIDING the Investigation
Part 3: Goldfish Behavior

1. **Focus question: What do goldfish do?**
   Call the first group of students to the center. Show them the square plastic tube you will put in the aquarium to make a fish tunnel. Ask,
   - ➤ *If we put a tunnel in the aquarium, what do you think the fish will do?*
   - ➤ *What have you seen them do when other things are added to the aquarium?*

   Have students explain their predictions, based on past observations of fish behavior.

2. **Put the tunnel in the aquarium**
   After students have had a few minutes to observe the aquarium, place the fish tunnel in the aquarium on the side away from the plants. Ask students to observe and describe what the goldfish do when the tunnel is first put into the aquarium, and after the fish have become accustomed to its presence.
   - ➤ *Did the fish seem frightened when the tunnel was placed in the aquarium?*
   - ➤ *What movements did the fish make to give you a clue that they might be afraid of this new object?*
   - ➤ *What did the fish do after they got used to the tunnel?*
   - ➤ *Do any of the fish swim through the tunnel?* **Next to** *the tunnel or* **above** *it?*

3. **Make a paper aquarium**
   Give each student in the group a *Paper Aquarium* sheet and a pair of scissors to make a three-dimensional model. If you have precut the aquarium pieces, pass them out one at a time as you guide students through the construction.

   a. Cut on the solid line labeled "1."

   b. Cut on the solid line next to the plants.

   c. Fold the paper on the dotted line so the plants picture will stand up on its own.

   Students could also cut slits between the plants so their fish can swim "through" the plants.

FOCUS QUESTION
What do goldfish do?

Say it → See it → Hear it → Write it → **New Word**

CROSSCUTTING CONCEPTS
Cause and effect

Materials for Step 2
- *Goldfish aquariums with plants*
- *Tunnels*

TEACHING NOTE
As students offer their observations, add any new or important vocabulary to the class word wall.

Materials for Step 3
- *Paper Aquarium sheets*
- *Tape*
- *Scissors*

Animals Two by Two Module—FOSS Next Generation

INVESTIGATION 1 – *Goldfish and Guppies*

SCIENCE AND ENGINEERING PRACTICES

Developing and using models

Analyzing and interpreting data

CROSSCUTTING CONCEPTS

Systems and system models

---

d. Cut out the tunnel on the solid line. Fold the paper in the same direction on each dotted line and tape the ends together to form a square tunnel that looks like the one in the goldfish aquarium.

e. Cut out the fish and fold back the flap so the fish can stand.

4. **Have a sense-making discussion**

Have students observe the aquarium and act out what they see with their paper aquariums. Ask the following questions to guide students:

➤ *Where are the fish swimming? Are they swimming near the surface of the water, at the bottom, or in the middle? Show me where the fish are swimming by moving your paper fish in its paper aquarium.*

➤ *Do the fish ever swim through the tunnel?*

➤ *Show me where the fish would be if it were behind the plants.*

➤ *Can the fish swim through the plants?*

➤ *Show me where the fish would be if it were next to the tunnel.*

Continue to ask questions that develop students' spatial vocabulary, using words such as *above*, **below**, *next to*, *through*, *behind*, **in front of**, and so forth.

Then discuss the model students have been using. Ask,

➤ *What are the parts of our paper model aquarium?* [Space, tunnel, fish, plant.]

➤ *What is not a part of our paper model but is a part of the real aquarium?* [Water, food, plastic basin.]

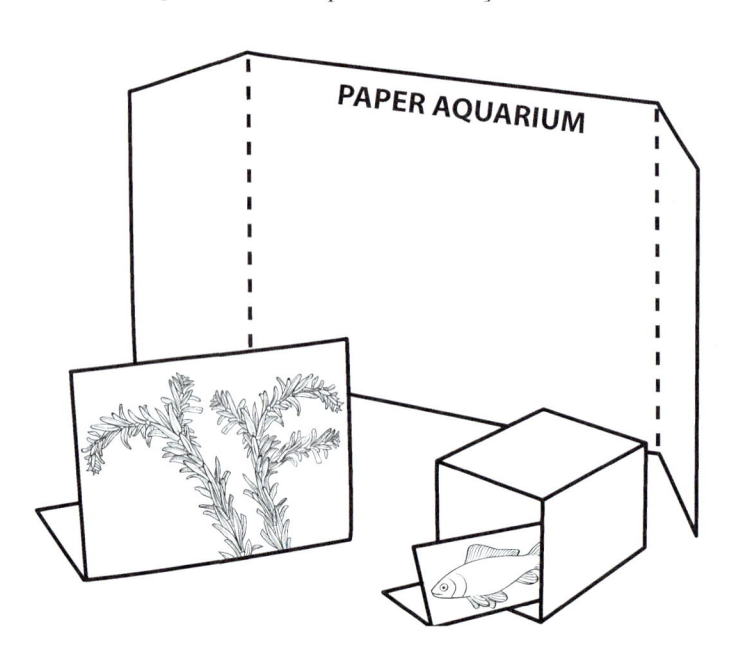

94

Full Option Science System

Part 3: Goldfish Behavior

5. **Review vocabulary**
   As students offer their observations, add any new or important vocabulary to the class word wall. Let students offer new vocabulary as needed.

6. **Answer the focus question**
   Ask each student to tell you (or a partner) one thing he or she observed today, before students leave the center.

   Tell students that you have a strip of paper with the focus question written on it. Describe how to glue the strip into the notebook before they answer the question. Ask them to return to their tables and work in their notebooks. For emergent writers, provide a sentence frame such as, I observe the goldfish _____ .

   Students can also draw pictures to show what they observed the goldfish do.

7. **Send the paper aquarium home**
   Give each student an envelope to take home all the pieces of his or her paper aquarium. Or students can keep the parts in their notebook. Prepare the center for the next group.

WRAP-UP/WARM-UP

8. **Share notebook entries**
   Conclude Part 3 or start Part 4 by having students share notebook entries. Ask students to open their science notebooks to the last entry. Read the focus question together.

   ➤ *What do goldfish do?*

   Ask students to pair up with a partner to
   - share their answers to the focus question;
   - explain their drawings.

   Students can also pretend they are goldfish and act out the different behaviors they observed.

---

above
below
in front of
next to
tunnel

Materials for Step 7
- *Envelopes*

FOCUS CHART

What do goldfish do?

Goldfish swim all around. They find out about things with their nose, mouth, and body.

TEACHING NOTE

See the **Home/School Connection** for Investigation 1 at the end of the Interdisciplinary Extensions section. This is a good time to send it home with students.

Animals Two by Two Module—FOSS Next Generation

INVESTIGATION 1 – *Goldfish and Guppies*

MATERIALS for
Part 4: *Comparing Guppies to Goldfish*

For each student
1 *FOSS Science Resources: Animals Two by Two*
- "Fish Same and Different"
- "Fish Live in Many Places"

For the class
1 Goldfish aquarium
1 Guppy aquarium (See Step 3, Getting Ready.)
❏ 1 Teacher master 9, *Center Instructions—Comparing Guppies to Goldfish*
1 Big book, *FOSS Science Resources: Animals Two by Two*

For assessment
- *Assessment Checklists* 1 and 2

★ Supplied by the teacher. ❏ Use the duplication master to make copies.

No. 9—Teacher Master

96 Full Option Science System

Part 4: Comparing Guppies to Goldfish

GETTING READY for
Part 4: Comparing Guppies to Goldfish

1. **Schedule the investigation**
   Each group of six to ten students will need about 15 minutes to observe the guppies and compare them with the goldfish. Plan 15 minutes for students to write or draw in their notebooks. Plan another session or two (30 minutes total) for the readings.

2. **Preview Part 4**
   Students compare the structures and behaviors of guppies to those of goldfish, and identify the guppies by gender. The focus questions are **How are guppies and goldfish different?** and **How are they the same?**

3. **Set up the guppy aquarium**
   If you have been hiding your guppy aquarium, bring it out for observation now. If you have postponed setting it up, get six feeder guppies—two male and four female—and follow the setup guidelines in Getting Ready for Part 1.

4. **Plan to read *Science Resources*: "Fish Same and Different" and "Fish Live in Many Places"**
   Plan to read "Fish Same and Different" and "Fish Live in Many Places" during a reading period after completing this part.

5. **Plan assessment for Part 4**
   Plan to observe students as they compare guppies to goldfish and to review their notebook entries.

   What to Look For

   - Students understand that fish are animals that have basic needs—food to live and grow, water with oxygen, and space. Animals live in places where their needs are met. (LS1.A: Structure and function; ESS3.A: Natural resources.)

   - Students use firsthand observations and readings to compare different kinds of fish structures and behaviors. (Asking questions; obtaining, evaluating, and communicating information; patterns; structure and function.)

Animals Two by Two Module—FOSS Next Generation

INVESTIGATION 1 – Goldfish and Guppies

FOCUS QUESTIONS

How are guppies and goldfish different?
How are they the same?

Materials for Step 1
- Goldfish aquarium
- Guppy aquarium

CROSSCUTTING CONCEPTS

Patterns
Structure and function

New Word: Say it, See it, Hear it, Write it

color
compare
different
female
guppy
male
same

TEACHING NOTE

Go to FOSSweb for Teacher Resources and look for the Science and Engineering Practices—Grade K chapter for details on how to engage kindergartners with the practice of asking questions.

SCIENCE AND ENGINEERING PRACTICES

Asking questions

GUIDING the Investigation
Part 4: Comparing Guppies to Goldfish

1. **Introduce the guppies**
   Send a group of students to the fish center to extend their understanding of "fishness" by introducing a new kind of fish. Introduce them to the aquarium with the guppies. Most students probably already know that these, too, are fish and might know they are called guppies. Let them observe unguided for a few minutes.

2. **Guide observations and comparisons**
   Ask students to **compare** the guppies to the goldfish. Discuss how they are the **same** and how they are **different**. Encourage students to ask questions.

   ➤ Do **guppies** and goldfish have the same body parts?

   ➤ Which kind of fish is bigger?

   ➤ Do they move the same way?

   ➤ Are they the same **color**?

   ➤ Will they eat the same kind of food? Try feeding them.

   ➤ What might happen if you put the tunnel in with the guppies? Will they react in the same way that the goldfish did?

   ➤ Some of the guppies are **males** (boys), and some are **females** (girls). Can you guess which is which? Why do you think so? [Males are generally smaller and have colored spots.]

   ➤ If someone showed you an animal that you had never seen before, how would you know if it was a fish? What would you look for?

   ➤ What questions do you have about guppies?

3. **Review vocabulary**
   As students offer their observations, add any new or important vocabulary to the word wall. Let students be the guides—confirm the words they use and offer new vocabulary as needed.

4. **Have a sense-making discussion**
   Sense-making discussions can be conducted in small groups as part of center time or as a whole class once all students have completed the center activity. The intent is for students to have a sense-making discussion after working with materials and before answering the focus question.

   Ask the following questions to guide students.

   ➤ What questions do you have about guppies?

   ➤ How are guppies and goldfish different? How are they the same?

98   Full Option Science System

Part 4: Comparing Guppies to Goldfish

Use a graphic organizer, such as a double bubble map shown here, to help students to organize the class observations. Demonstrate how to add one piece of information to the organizer in the form of a think aloud. Then ask students to provide more information that you can record.

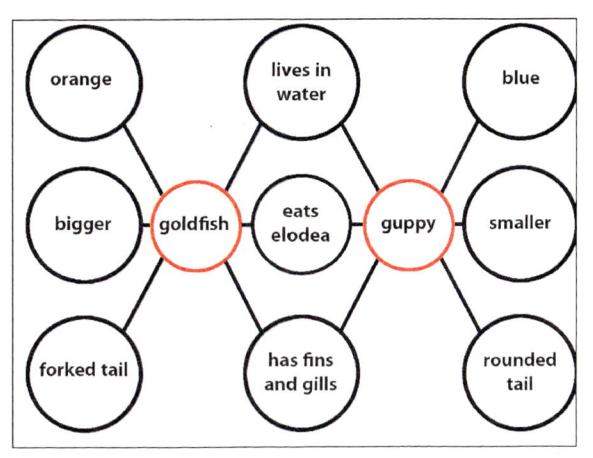

5. **Focus questions: How are guppies and goldfish different? How are they the same?**
   Write the focus questions on the chart as you read them aloud.

   ➤ *How are guppies and goldfish different? How are they the same?*

   Tell students that you have a strip of paper with the focus questions written on it. Describe how to glue the strip into the notebook before they answer the questions.

   When they return to their tables, they should answer the focus questions in their notebooks by drawing or describing in words and simple phrases, or by dictation. Ask them to include one or two things the fish have in common and one thing that is different.

   Ask each student to tell you one way that guppies and goldfish are either the same or different, before students leave the center.

6. **Call the next group**
   When one group has completed making observations, call another group to the center.

Male guppy

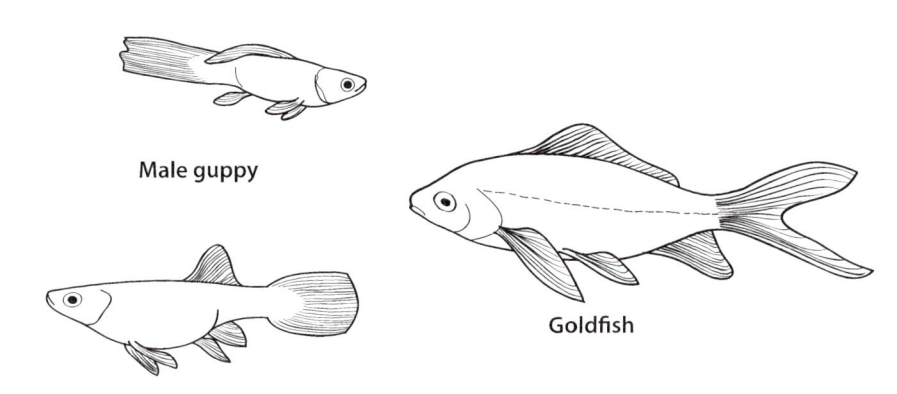

Female guppy

Goldfish

EL NOTE

Use a graphic organizer such as a Venn diagram, double bubble map, or box and T-table to help students compare and contrast.

SCIENCE AND ENGINEERING PRACTICES
Constructing explanations

FOCUS CHART

How are guppies and goldfish the same?

Both swim.

Both need water and food.

Both have fins, eyes, mouth, scales, and tail.

FOCUS CHART

How are guppies and goldfish different?

Guppies	Goldfish
small	big
long	oval
one tail	forked tail
boys spotted	girls and boys look same

Animals Two by Two Module—FOSS Next Generation

INVESTIGATION 1 – Goldfish and Guppies

READING in Science Resources

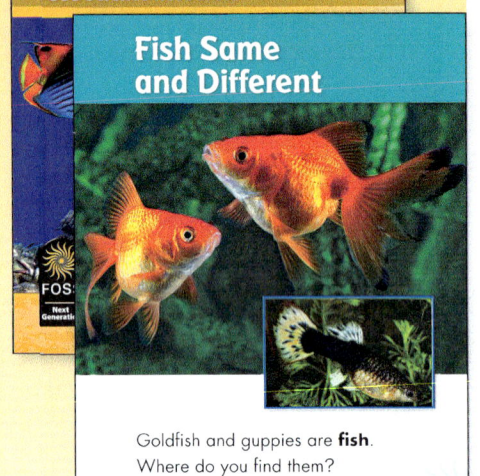

7. Preview the book cover

Gather students on the rug and show the big book (or student book version). Ask them to identify the front cover, the back cover, and the title page. Read the title, and give students a few minutes to look at and discuss the photograph on the front cover.

Introduce the table of contents and demonstrate how it is used to organize and locate information. Tell students that this book might provide answers to some of their questions about fish and other animals they will be exploring in class.

Read the title of the article, "Fish Same and Different," and have students turn and talk to a neighbor about what they have learned so far about how fish are the same and how they are different. Tell students that you will read the article to them a few times.

First, students will preview the text by looking at the pictures and talking with their neighbor about what they notice that is the same and different about fish. Discuss any other text structures and features they notice such as the words in bold and question marks. Explain that the words in bold might be new to them and are important to understanding the text and illustrations. Tell them,

If a reader does not know what the word means, he or she can find out by looking at the glossary at the back of the book.

The question marks tell us that the author is asking us questions to help us think about what we are reading.

8. Read "Fish Same and Different"

Read the article aloud. Pause and allow students to answer the questions posed in the text and record any additional questions they have. Model how you might try to figure out the meaning of the word *oxygen*. Tell them,

*The text says, "They need **oxygen**, too." But I can't tell what oxygen is from the photographs or the other sentences. I'll look it up in the glossary.*

9. Discuss the reading

Discuss the generalizations that can be made about the animals. For instance,

➤ *What parts do all fish have?* [Gills, fins, scales, eyes, mouth, tail.]

➤ *How do fish differ from one another?* [Color, size, shape.]

Encourage students to use information from the article to help answer the questions.

ELA CONNECTION

These suggested strategies address the Common Core State Standards for ELA.

RI 1: Ask and answer questions about key details.

RI 3: Describe the connection between two ideas.

RI 4: Ask and answer questions about unknown words.

RI 5: Identify the front cover, back cover, and title page of a book.

RI 10: Actively engage in group reading activities with purpose and understanding.

SL 4: Describe with details.

CROSSCUTTING CONCEPTS

Patterns

Part 4: Comparing Guppies to Goldfish

10. Share information about photos

Flip through the pages of the article again. This time tell students you will share some of the interesting features of the animals.

Page 3. Fantail goldfish (top), Golden leopard delta guppy (bottom). Goldfish come in different shapes, sizes, and colors. This one is named after the shape of its tail. Goldfish are egg-layers. Size: 5-6 cm.

This guppy variety is raised only in fish tanks. You'll never find it in the wild. Guppies bear their young live. Size: 2.5 cm.

Page 4. Koi (top inset), Striped large-eye bream (bottom). Koi are all descendents of a species of black carp called maoi, raised in China nearly 2000 years ago. Koi are related to goldfish. Large size: 35-60 cm.

This bream is also called Yellow-spot Emperor. They live in the Indo-Pacific ocean near Micronesia. Striped large-eye bream can quickly "turn off" their stripes and spots to hide from predators. Size: 30 cm.

Page 5. Coral reef fish in saltwater tank (top), rainbow trout in freshwater stream (bottom).

See if your students can find these colorful reef fish. Their names are clues to their identity: Yellow tang, blue tang, clownfish, butterfly fish, damselfish, angelfish.

These rainbow trout live in streams and rivers. Another type of trout called steelheads make their way to the ocean where they live for 2-3 years. The adults will return to the freshwater stream where they hatched to lay eggs for the next generation of trout.

Page 6. Rainbow trout (top). The pink, green, and blue coloration gives this trout its name.

Page 7. Piranha (top), Goldfish (bottom). Piranha live in the rivers of the Amazon in South America. They have many sharp teeth for eating fish and other animals. The gill cover or slit is visible on both the piranha, and goldfish in the inset. Water enters the fish through the mouth where it passes over the gills and exits out the opening. The gills extract the needed oxygen from the water. Size: 20-45 cm.

> **TEACHING NOTE**
> You can draw the size of the animal on chart paper or use a meter tape to show students the size.

SCIENCE AND ENGINEERING PRACTICES

Asking questions

Obtaining, evaluating, and communicating information

Animals Two by Two Module—FOSS Next Generation

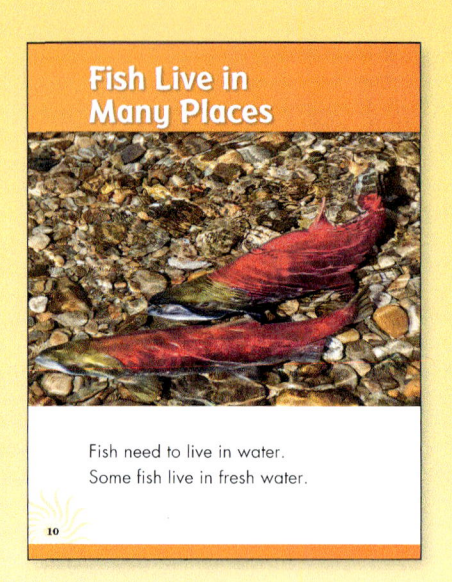

INVESTIGATION 1 – *Goldfish and Guppies*

Fish Live in Many Places

Fish need to live in water.
Some fish live in fresh water.

10

SCIENCE AND ENGINEERING PRACTICES

Asking questions

Obtaining, evaluating, and communicating information

CROSSCUTTING CONCEPTS

Patterns

ELA CONNECTION

These suggested strategies address the Common Core State Standards for ELA.

RI 2: Identify main topic and retell key details.

SL 2: Ask and answer questions about key details and request clarification.

SL 3: Ask and answer questions to seek help, information, or clarify.

11. Read "Fish Live in Many Places"

Introduce the next article and ask students to discuss what they think they will learn. Brainstorm a list of the places where students think fish live.

Read the article aloud. Pause and allow students to answer the questions posed in the text and record any additional questions students have. Encourage them to identify the structures of the fish.

12. Discuss the reading

Ask students to think–pair–share what they think was the important idea in this article. Confirm that the main idea is that fish have basic needs that are met by living in water. Have students retell the key details of the text using these questions.

➤ *What are the basic needs of fish?* [Water to live in and to get air from, food, space, shelter.]

➤ *Where do fish live?* [In marshes, the ocean and saltwater aquariums; fresh water in lakes, streams, ponds, and aquariums.]

Review the initial list of places fish live and add any new information gained from the article.

13. Share information about photos

Tell students you will review the article to share some of the interesting features of the animals. As you are discussing the photographs, encourage students to look for patterns of what fish need to survive.

Page 10. Spawning salmon. Salmon live their adult life in the ocean but return to the river or stream where they hatched to lay eggs for the next generation of salmon. The females go through a color change, turning bright red as they return to fresh water.

Page 11. Rainbow trout. Rainbow trout live in cold lakes and streams. They eat small fish, crustaceans, and insects. Size: 51–76 cm.

Page 12. Marlin (bottom). A marlin is a type of billfish, having a long spear-like snout or bill. They are fast swimmers, reaching speeds of 110 kilometers per mile (68 mph). The Atlantic blue marlin is one of the larger species, weighing in at 818 kilograms (1,800 lbs.). Size: 6 m.

Page 14. Huge saltwater tank at an aquarium. A variety of ocean fish can live together in this saltwater habitat. Have your students spot the two bat rays. These fish have a head and "wings" like a bat. The "wings" of a bat ray are fins.

Part 4: Comparing Guppies to Goldfish

Page 18. Redfin perch (bottom). Redfin perch live in slow-moving water such as lakes and swamps. They prefer areas with plants and rocks for shelter, and will eat small fish, insects, and worms.

Page 19. Plecostomus. A plecostomus is also called a sucker fish. It eats algae off rocks and the sides of tanks. Size: 7.5–60 cm.

To help students focus on the similarities and differences of different animals, start a content grid with fish. Continue adding rows to the chart as the class explores new organisms.

Animal	Parts	What they do	Where they live	What they need to live
fish	head tail fins mouth scales	swim eat explore	water (ponds, lakes, rivers, ocean, aquariums)	food clean water space shelter
birds				

WRAP-UP/WARM-UP

14. Share notebook entries

Conclude Part 4 or start Part 5 by having students share notebook entries. Ask students to open their science notebooks to the last entry. Read the focus questions together.

➤ *How are guppies and goldfish different? How are they the same?*

Ask students to pair up with a partner to

- share their answers to the focus questions;
- explain their drawings.

Model for students how to critique their notebook entries. Make an entry on the class notebook or chart paper (or use an anonymous student sample from a previous year) and have a volunteer say one thing that is good, e.g., the picture clearly shows how the fish are different.

Then ask for a volunteer to say one thing that would make the entry better, e.g., add labels to the drawing. Have students go back to their own notebook drawings and share with a partner what they like about their own entry and what they think they can do to make it better. Give students a few minutes to revise their entries.

CROSSCUTTING CONCEPTS

Patterns

ELA CONNECTION

This suggested strategy addresses the Common Core State Standards for ELA.

W 5: Strengthen writing.

INVESTIGATION 1 – *Goldfish and Guppies*

MATERIALS *for*
Part 5: *Comparing Schoolyard Birds*

For each student
- 2 Toilet-paper tubes ★
- 1 String, 80 cm ★
- 1 *Bird Outline* ★
- 1 *FOSS Science Resources: Animals Two by Two*
  - "Birds Outdoors"

For the class
- 1 Clipboard ★
- 3 Sheets of white paper ★
- 1 Hole punch ★
- 1 Stapler ★
- 1 Bird guide ★
- • Chart paper ★
- 1 Large sketch of tree (See Step 7 of Getting Ready.) ★
- 1 Set of bird silhouettes
  - 1 American crow
  - 3 American robin
  - 6 House sparrow
- • Computer with Internet access ★
- • Projection system ★
- ❏ 1 Teacher master 10, *Bird Outline*
- 1 Big book, *FOSS Science Resources: Animals Two by Two*

For assessment
- • *Assessment Checklists* 1, 2, and 3

★ Supplied by the teacher.   ❏ Use the duplication master to make copies.

No. 10—Teacher Master

104   Full Option Science System

Part 5: Comparing Schoolyard Birds

GETTING READY *for*
Part 5: Comparing Schoolyard Birds

1. **Schedule the investigation**
   This part will take three whole-class sessions for active investigation. Plan 15 minutes for students to write or draw in their notebooks. In addition, plan one reading session of 15 minutes.

2. **Preview Part 5**
   Students go bird watching to observe and compare the structures and behaviors of two types of common schoolyard birds. The focus question is **What birds visit our schoolyard?**

3. **Select your outdoor site**
   Get to know the birds in your schoolyard. Walk around your schoolyard and listen and look for birds on the ground, in bushes or trees, and on fences or wires. Determine the route you will take with students when they go on three bird walks. You might walk around the school on a zig-zag course across a field or on an established path. Avoid paths with lots of tree roots. Ideally, there will be a few dry places along the route to stop, sit, and look up for birds—asphalt works well for this, or students can lean against a wall. If your schoolyard has trees, make sure you plan to stop for a few minutes to scan the branches and listen carefully.

4. **Plan for time of day**
   Birds will be most active early in the morning. Try to schedule your bird walks accordingly. Also, avoid times of day when other students are outside at recess.

5. **Research local birds**
   Find out what common birds are in your schoolyard and in your community. Check with a naturalist at a local park or science center, or go online to a local bird-watchers' site. You can find links to good bird-identification sites by going to the Resources by Investigation section of FOSSweb.

   In Step 18, you will project the photos of a few of the birds that students observe in the schoolyard. Practice how to search and find these birds and what information is available. There will be still photos, audio recordings, and sometimes videos.

6. **Plan for additional help**
   It is helpful to have an assistant (either an adult or an older student) accompany you and the group on the bird walks.

7. **Display bird silhouettes**
   The kit contains a set of life-sized bird silhouettes showing the relative bird size from the tip of the bill to the tip of the tail. The

> **TEACHING NOTE**
> You might want to extend the schoolyard bird exploration throughout the entire module. Check out the Interdisciplinary Extensions for more bird watching suggestions.

▶ **NOTE**
Enlarging your learning space to include the schoolyard and the local neighborhood requires implementing some new teaching methods and learning behaviors. For a more detailed discussion of these methods, see the Taking FOSS Outdoors chapter in *Teacher Resources*.

Animals Two by Two Module—FOSS Next Generation

INVESTIGATION 1 – *Goldfish and Guppies*

TEACHING NOTE

Using binoculars is challenging and distracting for many young students. You might see that, even with the paper binoculars, students hold on to them but do not use them. The binoculars still serve an important function. They focus the investigation by acting as a reminder that students are on a bird walk and not at recess.

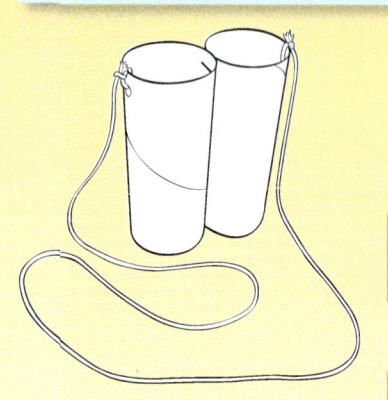

three birds in the silhouette set are a house sparrow (about 12.5 cm), an American robin (about 25 cm), and an American crow (43 cm). There are six sparrows, three robins, and one crow for a total of ten birds. Display nine of the bird silhouettes around the classroom when students are out of the room (hold on to one of the sparrow silhouettes for the introduction in Step 2). Make the birds stand on the ground or perch on a fence, wire, or tree limb. Sketch a tree with many branches on a piece of chart paper and place a few birds in the tree.

8. **Make binoculars**
   Place two paper tubes side by side and make sure the tubes are parallel. Staple them together at the top and bottom edges. Look through the binoculars to make sure you see one "circle-of-view" and that the tubes aren't askew. If they are askew, simply take one of the staples out, realign the tubes, and staple again. Punch a hole near the ends of both rolls and tie a string between the two holes. Students can then wear their binoculars like a necklace.

9. **Check the site**
   It is always a good idea to check the outdoor site on the morning of an outdoor activity. Check for any distracting or unsafe items where students will be working.

10. **Plan to read *Science Resources*: "Birds Outdoors"**
    Plan to read "Birds Outdoors" during a reading period.

11. **Plan assessment for Part 5**
    Plan to observe students as they study local birds. This is a good time to assess students engagement with crosscutting concepts. This is a good time to reflect on student engagement with the crosscutting concepts of patterns, cause and effect, systems and system models, and structure and function.

    What to Look For

    - Students know that animals have external parts that help them to live, get food, and meet their needs. (LS1.A: Structure and function; LS1.C: Organization for matter and energy flow in organisms; patterns.)

    - Students describe observations of the structures of birds, record information by writing and labeling pictures, and share their notebook entries with others. (Analyzing and interpreting data.)

Part 5: Comparing Schoolyard Birds

GUIDING *the Investigation*
Part 5: Comparing Schoolyard Birds

1. **Move the focus to birds**
   Gather students at the rug. Display a picture of fish from the *FOSS Science Resources* big book and ask,
   ➤ *What tells you that this animal is a fish?*
   ➤ *What do fish need to live and grow?*

   Tell students that now they will observe another type of animal and determine how it lives and what it needs to grow. Give some clues to see if they can guess the identity.

   This animal lives outside. It has two feet. It has two eyes. It has **wings**. *It flies.*

   After students identify the animal as a bird, ask,
   ➤ *Has anyone seen a* **bird** *in our schoolyard? Where?*

2. **Introduce bird watching**
   Tell students that they will go on several bird walks to observe birds in the schoolyard. Say,

   Before we go outdoors, we need to practice how to act so that we don't scare the birds away when we're on our bird walk. We need to walk slowly, silently, and with quiet footsteps. When we see a bird, we should point to it so that others will see it too. If we hear a bird, we can point and gently tug on our ear.

   Hold up the sparrow bird silhouette. Explain to students that you want them to walk around the room and practice going on a bird walk. Tell them there are nine birds they should look for in the room.

   If students are breaking the bird-walk protocol, stop the action, refocus on the proper expectations, and let them try again.

3. **Review bird watching**
   Ask students where they saw the birds. Emphasize that birds are seen in the air, on the ground, in trees, on wires, and sometimes in puddles or ponds.

4. **Discuss outdoor safety rules**
   Go over the rules listed on the *Outdoor Safety* poster. Remind students to behave appropriately so they have a good bird walk and everyone stays safe.

FOCUS QUESTION
What birds visit our schoolyard?

EL NOTE
Use gestures to describe the part of the bird if necessary.

Say it / See it / Hear it / Write it — **New Word**

Materials for Step 2
- Bird silhouettes
- Sketch of tree

INVESTIGATION 1 – *Goldfish and Guppies*

Materials for Step 5
- Clipboard with paper
- Pen
- Binoculars

TEACHING NOTE

The focus on the first bird walk is to simply observe and get excited about birds. It is not necessary to name the birds. If students can name the bird they may share, but this is not the point.

SCIENCE AND ENGINEERING PRACTICES

Planning and carrying out investigations

Materials for Step 7
- Projection system

EL NOTE

Write the names of the body parts on a diagram of a bird. Allow time for students to practice saying the words.

Say it • See it • Hear it • Write it
New Word

5. Go outdoors

After students are appropriately dressed for the outdoors, tell them that they are going to use their senses as a tool of observation to identify birds. Introduce another tool, the faux binoculars, and demonstrate how to use them, and then distribute a pair to each student. Head outside to the spot at the beginning of your selected route. Stand silently for a minute and look around for birds. If necessary, remind students how to move as quietly as possible and to point to birds (or bird sounds) to alert classmates.

Walk your planned route, stopping whenever you see or hear a bird. If you see a bird, write it on your bird list. On this first bird walk, this list is just for your information.

Ask questions, including

➤ *Where is the bird?*

➤ *What size is the bird (compared to a sparrow, robin, or crow)?*

➤ *What color is the bird?*

➤ *What questions do you have about birds?*

6. Return to class

Store the binoculars so that the strings don't get tangled. Gather students at the rug and ask,

➤ *Did you see birds flying in the air? Where else did you see them?*

➤ *What did you notice about the birds?*

BREAKPOINT

7. Introduce bird structures

Gather students at the rug. Project a photo or drawing of a bird seen on the first walk and say,

On our last bird walk, we saw this bird.

➤ *What do you notice about the bird?*

➤ *What parts or* **structures** *does the bird have?*

If students need encouragement, ask them to focus on any observable body parts: head, **bill**, crown, belly, back, neck, tail, wing, and **feathers**. As they describe the parts, point to them. Students should compare the bird body parts to their own body parts.

108 Full Option Science System

Part 5: Comparing Schoolyard Birds

8. Focus question: What birds visit our schoolyard?
Write the focus question on the chart as you read it aloud.

➤ *What birds visit our schoolyard?*

Explain that students will be going outdoors to bird watch, and their challenge is to see if they can find the same bird they found yesterday. When they find the bird, or any bird if viewing is limited, they should observe it and see what it is doing.

9. Go outdoors
Explain that today students will go on another bird walk. Take a clipboard and pen to record bird sightings. Ask students to remind you of the proper bird watching behavior. Redistribute the binoculars and head outdoors to an outdoor gathering spot.

Start with a silent minute of standing and observing. Then begin the slow bird walk around the schoolyard. Remember to stop every so often to look and listen. Stop every time you see a bird of any kind, particularly the common schoolyard bird that students are challenged to observe.

10. Discuss bird sightings
When students get a good look at a bird, ask them to describe it.

➤ *What size is the bird (compared to a sparrow, robin, or crow)?*

➤ *What color is the bird?*

➤ *How big is the bill? Can you make the shape of the bird's bill with your fingers?*

➤ *How does the bird hold its wings when it is standing?*

➤ *How does the bird **fly**?*

When you see a bird flying, stop and ask students to observe how the bird is moving its wings. Ask students to imitate the bird with their "wings." Ask them to repeat this as they see other birds.

11. Have a sense-making discussion
Return to class. Have students give you their binoculars and join you at the rug with the bird list. Use these questions to guide the discussion.

➤ *Did we find the bird we were looking for in the schoolyard?*

➤ *How many birds did we find?*

➤ *What were they doing?*

➤ *What other birds did we see and what were they doing?*

➤ *What questions do you have about birds in our schoolyard?*

Materials for Steps 8–9
- *Bird guide*
- *Clipboard with paper*
- *Pen*
- *Binoculars*

SCIENCE AND ENGINEERING PRACTICES

Asking questions

Planning and carrying out investigations

Analyzing and interpreting data

CROSSCUTTING CONCEPTS

Systems and system models

Structure and function

Animals Two by Two Module—FOSS Next Generation

INVESTIGATION 1 – *Goldfish and Guppies*

bill
bird
feather
fly
structure
wing

12. Review vocabulary
As students offer their observations, add any new or important vocabulary to the class word wall. Let students be the guides—acknowledge the words they use and offer new vocabulary as needed.

13. Label bird structures
Distribute copies of teacher master 10, *Bird Outline*. Have students draw in any missing structures on the bird outline, and label the structures they know. Students should put the sheet in their notebook and write a few words about the birds they observed in the schoolyard.

POSSIBLE BREAKPOINT

14. Discuss relative positions
Display the drawing of the tree you created in the first bird watching session. Have students shut their eyes while you place a few sparrows on the tree. Have them keep their eyes shut. Tell them,

Imagine you are near a big, tall tree in the schoolyard. You see a bird on the tree, but nobody else can see it. You need to describe where the bird is on the tree to help your classmates find it.

Have students look at the tree and with their rug partner, take turns describing where on the tree the birds are hiding. If they do not yet know their right from their left, they can point left or right. Help them describe location:

- The bird is on this side <point to the left> on the second branch from the bottom.
- The bird is on the trunk just above the lowest branch.
- The bird is at one of the very top branches on this side <point to the right>.

Using one robin and a sparrow, position the sparrow in different locations in relation to the robin, and ask students to describe the position of the sparrow. Give students practice observing and describing the location of one bird in relation to the other.

Materials for Step 14
- *Sketch of tree*
- *House sparrow silhouettes*
- *Robin silhouette*
- *Transparent tape*

15. Go outdoors
Explain that today students will go on another bird walk. Their challenge is to observe a different kind of bird. They should find out where it goes and what it does. Ask students to suggest a place in the schoolyard where they might observe different birds.

Part 5: Comparing Schoolyard Birds

Ask students to remind you of the proper bird watching behavior. Redistribute the binoculars, and head outdoors to a gathering spot. Ask students,

➤ *What birds do you think we will see on this walk?*

Start with a silent minute of standing and observing. Begin the slow bird walk around the schoolyard. Remember to stop and observe every so often and certainly every time you see a bird.

16. **Discuss bird movement**
    Search for birds in trees and emphasize describing where the bird is on the tree. Students can pair up and describe to their partner where the bird is. Continue to describe the birds' features and movement when you see them. Ask questions to inspire observations:

    ➤ *What size is the bird (compared to a sparrow, robin, or crow)?*

    ➤ *What color is the bird?*

    ➤ *Where is the bird (on the ground, in a tree, flying)?*

    ➤ *What is the bird doing?*

    ➤ *How does the bird fly?*

    ➤ *What sounds does the bird make?*

    ➤ *What questions do you have about birds?*

17. **Return to class**
    Have students give you their binoculars and join you at the rug. Tell students that they will be drawing in their notebook one of the birds they observed in the schoolyard. If necessary, review the expectations for a scientific drawing. Students might want to label some parts. Make new words available on the word wall.

18. **Project images of birds**
    Project pictures of several of the birds that students observed in the schoolyard. Go to the Resources by Investigation section on FOSSweb to find a link to the recommended bird-identification sites. If available, play the audio links so students can hear the bird calls, and play any videos of the bird's behavior.

19. **Review vocabulary**
    As students offer their observations, add any new or important vocabulary to the word wall. Let students be the guides—acknowledge the words they use, and offer new vocabulary as needed.

SCIENCE AND ENGINEERING PRACTICES
Asking questions
Planning and carrying out investigations

CROSSCUTTING CONCEPTS
Patterns

Animals Two by Two Module—FOSS Next Generation

INVESTIGATION 1 – *Goldfish and Guppies*

SCIENCE AND ENGINEERING PRACTICES

Asking questions
Constructing explanations

FOCUS CHART

What birds visit our schoolyard?

Robins look for food on the lawn.

Doves sit on the wire and call.

20. Answer the focus question

Ask students to turn and talk to a neighbor about the birds that visit the schoolyard the most. Ask them to think about the structures those birds have and how they use those structures. If there is time, have students share with a partner what they learned about those birds. Then, use a whip-around strategy and have each person say one thing that they learned about the birds.

Restate the focus question, and have the class read it aloud together.

➤ *What birds visit our schoolyard?*

Tell students you have a strip of paper with the question written on it. Review how to glue the strip into the notebook.

Ask students to answer the focus question in pictures and/or words.

For emergent writers, provide a sentence frame such as, I observed _____ in our schoolyard. It has _____.

If possible, have students dictate what they learned about the different birds that visit the schoolyard and what questions they have.

112　　　　　　　　　　　　　　　　　　　　　　　　　**Full Option Science System**

Part 5: Comparing Schoolyard Birds

READING *in Science Resources*

21. Read "Birds Outdoors"
Begin by introducing the title of the article, "Birds Outdoors." Tell students that this article will help them understand more about the similarities and differences of birds in the wild. Photos and text might answer some of their questions about where birds live and what they need to survive.

Read the article aloud, using the strategies that will be most effective for your class. Pause to discuss key points in the article, to compare the photos, and to respond to the questions. This is also a good opportunity to discuss the relationship between the photographs and the text. Ask students how the photographs help them understand what the author is saying about birds.

22. Discuss the reading
Discuss the article, using these questions as a guide.

➤ Where do you find birds outdoors?

➤ What do birds eat?

➤ What else do they need to live?

➤ Birds have wings. What other body parts do they have?

➤ How are birds different? How are they the same?

Add birds to the content grid (started with the reading in Part 4) and have students use information from the text to help fill in the columns.

23. Share information about photos
Tell students you will review the article to share some of the interesting features of the animals. As you are discussing the photographs, encourage students to look for patterns of what birds need to survive.

Page 20. Canadian geese (top), Black-capped chickadee (inset). The Canada goose is recognized by its black head and neck with white patches on the cheeks. Some of these geese migrate in winter from Canada to the warmer areas of the southern United States. Size: 76–110 cm.

The Black-capped chickadee is a common bird at birdfeeders and in bushes and trees. It is named after the black cap on its head and the call it makes: chick-a-dee-dee-dee. Size: 13 cm.

Page 21. Eastern bluebird (bottom). This is a male bluebird because of its royal blue head and back. The females have a

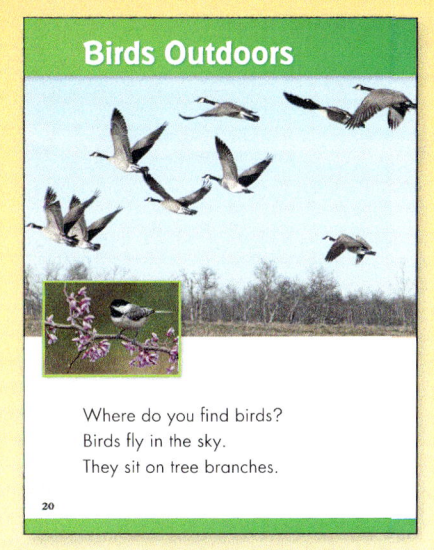

Birds Outdoors

Where do you find birds?
Birds fly in the sky.
They sit on tree branches.

20

ELA CONNECTION

These suggested strategies address the Common Core State Standards for ELA.

RI 1: Ask and answer questions about key details.

RI 2: Identify main topic and retell key details.

RI 7: Describe the relationship between illustrations and the text.

RI 10: Actively engage in group reading activities with purpose and understanding.

W 8: Gather information to answer a question.

SCIENCE AND ENGINEERING PRACTICES
Obtaining, evaluating, and communicating information

CROSSCUTTING CONCEPTS
Patterns

Animals Two by Two Module—FOSS Next Generation

INVESTIGATION 1 – *Goldfish and Guppies*

brownish head and back. They perch on posts, wires, and low branches to pounce on insects.

Page 22. Great horned owls (left inset), Herring gull and sea star (right inset), bald eagle and salmon (bottom). The feather tufts on a great horned owl's head look like horns. The babies are fluffy when young and will develop the tufts as they become adults. This gull will scavenge the shore looking for sea stars and crabs to eat.

A bald eagle will use its powerful talons to grab a salmon from below the surface of the water.

Page 24. Common loon (top). The loon has various calls, hoots, yodels, and wail sounds that carry across a lake. The black and white patterns on their neck, back, and their red eyes identify them as a loon.

Page 25. Northern cardinal (left), cedar waxwing (right). This male northern cardinal sits low in trees and shrubs. They are common at birdfeeders and tend to perch like this, hunched over with tail pointed straight down.

Cedar waxwings will swallow berries whole. They also fly expertly over rivers to catch insects.

Page 26. Yellow warbler (left), bald eagle (top right). Yellow warblers have a whistle-like call that sounds like "sweet sweet sweet I'm so sweet." They build their nests in small trees and thickets for protection.

Bald eagles build their huge nest of small branches high on top of dead trees or powerline crossbars.

WRAP-UP

24. Share notebook entries

Conclude Part 5 by having students talk with a partner about the guiding questions for the investigation. They should use their notebooks as a reference. After sharing with a partner, ask for volunteers to talk about their ideas.

➤ *What do fish need to live and grow?*

➤ *What do birds need to live and grow?*

➤ *What is the same about what birds and fish need; what is different?*

➤ *What did you find out about fish or birds that was most interesting to you?*

DISCIPLINARY CORE IDEAS

LS1.A: Structure and function

LS1.C: Organization for matter and energy flow in organisms

ESS2.E: Biogeology

ESS3.A: Natural resources

ELA CONNECTION

This suggested strategy addresses the Common Core State Standards for ELA.

SL 1: Participate in collaborative conversations.

Interdisciplinary Extensions

INTERDISCIPLINARY EXTENSIONS

Language Extensions

- **Write a story**
  Write a class or group story titled "A Day in the Life of a Fish."

- **Write a "facts about fish" book**
  Cut paper into the shape of a fish. Use it to record ideas from students about what fish need to live, how to care for fish, and facts they know about fish from things they've read or field trips they've been on. Have students draw on the fish-shaped pages, cut pictures from catalogs and magazines, or find pictures online to illustrate their book.

Math Extensions

- **Count the fish in the tanks**
  Ask students to count the fish in the tanks. Use paper fish or goldfish crackers to represent male guppies, female guppies, and goldfish.

- **Add and subtract with fish**
  Have students use paper fish or goldfish crackers to work out simple addition and subtraction problems.

Art Extension

- **Make your room an aquarium**
  Make large, stuffed paper fish to hang from the ceiling in your room. Have students paint them. Plan a mural on a large bulletin board. Hang paper plants in the environment as well.

Science Extensions

- **Train the goldfish**
  Goldfish are trainable. See if you can train them to come to one end of the aquarium, moving toward a brightly colored object dangled in the aquarium just before feeding time.

TEACHING NOTE

Refer to the teacher resources on FOSSweb for a list of appropriate trade books that relate to this module.

TEACHING NOTE

Encourage students to use the Science and Engineering Careers Database on FOSSweb.

TEACHING NOTE

Review the online activities for students on FOSSweb for module-specific science extensions.

INVESTIGATION 1 – *Goldfish and Guppies*

- **Obtain other fish food**
  If students are interested in what fish eat, get another food such as tubifex worms, live or dried brine shrimp, bits of fish such as tuna, or finely chopped green leafy plants. Be sure to keep the water clean if you add these foods. It might require changing the water more frequently.

Environmental Literacy Extensions

- **Identify animals that eat fish**
  Ask students to use the computer to find pictures of animals that eat fish for food. Bears, seals, orcas, cats, and people are just a few.

- **Play a camouflage game**
  Prepare a camouflage pond and stock it with camouflaged fish.
  a. Cut off the top third of a large brown-paper grocery bag. Fold the edge over so the bag will stand up. This is your pond.
  b. Line the inside bottom of the bag with newspaper and glue it down. Use the want-ads section for this activity.
  c. Tear strips of newspaper (2 cm wide) and toss them in the bag to create a "seaweed" forest.
  d. Cut fish from newspaper and brightly colored construction paper—about twice as many newspaper fish as colored fish. Mix the fish in with the newspaper strips.

  Invite students to go fishing. Tell them to pretend that they are a great big fish looking for a smaller fish for lunch. Pass the bag around and ask students to look in the pond and "find a fish to eat." When everyone has had a turn, have them compare their fish. (In most cases, all will have chosen a brightly colored fish.)

  Show students a newspaper fish. Tell them that there were many more newspaper fish in the pond than colored fish. Ask them to explain why the newspaper fish didn't get eaten for lunch. [Because they looked just like the seaweed, so they were camouflaged; they were hard to see.]

- **Set up a birdfeeder**
  Have students use the Internet to research different kinds of birdfeeders and bird food and find the feeder best suited to the schoolyard habitat. Work with the custodian to set up the feeder and manage it over time. Have students observe the feeder at different times of the day and keep a record of all the birds that visit the feeder.

Interdisciplinary Extensions

- **View video on peregrine falcons**
  Have students find out how scientists work to protect endangered birds. View "The Urban Habitat of Peregrine Falcons" (duration 6 minutes, chapter 5 of *Is This a House for Hermit Crab?* video). Students see scientists who work in the city to protect and provide roosts for peregrine falcons. A link to this video clip is in the Resources by Investigation section of FOSSweb.

  The video chapter shows how peregrine falcons live among the skyscrapers in the city. They were an endangered species for many years.

  Peregrine falcons make their homes on the ledges of tall buildings and bridges; the underside of bridges is a favorite place for falcons to raise their families; scientists band birds for identification purposes.

  After babies or hatchlings are banded, they get a complete checkup; through the efforts of courageous scientists, peregrine falcons make comfortable homes in our cities.

- **Take a field trip to a body of water**
  Plan a class field trip to a pond, lake, marsh, or beach. Focus the trip on the needs of the animals in the area. Contact a local environment education center and find out what field trips they offer. Check the regional resources on FOSSweb for possible organizations to contact.

Home/School Connection

Print or make copies of teacher master 11, *Home/School Connection* for Investigation 1, so each student has a picture of a fishbowl and a picture of a fish to color. Send it home at the end of Part 3.

At home, students attach both pictures, back to back, to the end of a drinking straw or pencil. They hold the straw between their palms with the pictures up, spin the straw back and forth, and watch the picture. An optical illusion makes the fish look as if it were in a bowl.

As an additional home/school connection, ask students to talk with their families about connections they have to fish or to birds. Do they like fish? Do they like birds? Do they have any artwork of fish or birds? Is there a traditional story about fish or birds?

Invite family members to the classroom to share their knowledge about, experience with, or cultural connections to fish and/or to birds.

TEACHING NOTE

Families can get more information about Home/School Connections on FOSSweb.

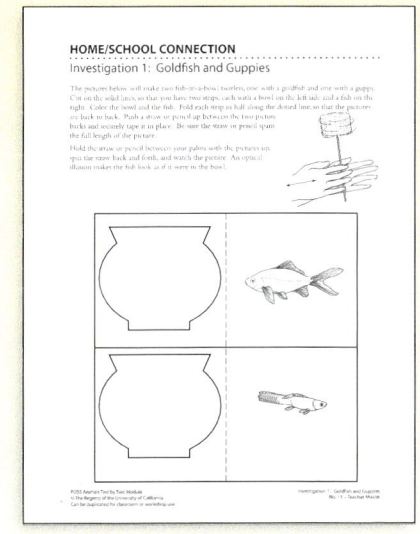

No. 11—Teacher Master

Animals Two by Two Module—FOSS Next Generation

117

INVESTIGATION 1 – Goldfish and Guppies

INVESTIGATION 2 – Water and Land Snails

Part 1 Observing Water Snails	130
Part 2 Shells	137
Part 3 Land Snails	142

Guiding question for phenomenon:
What do animals such as snails need to live and grow?

PURPOSE

Students have firsthand experiences with two related animals—water snails and land snails. Through observation and discussion, students gather information about snail structures and behaviors and how those characteristics relate to the needs of the animals. Students focus on one phenomenon: the snail's protective shell.

Content

- Different kinds of snails have some structures and behaviors that are the same and some that are different.
- Snails are animals and have basic needs—water, air, food, and space with shelter.
- There is great diversity among snails. Shells differ in size, shape, pattern, and texture.
- Snails have senses.

Practices

- Observe the structures and behaviors of land snails in a terrarium and water snails in an aquarium.
- Describe, compare, and communicate the similarities and differences of the two kinds of snails.

Science and Engineering Practices

- Asking questions
- Planning and carrying out investigations
- Analyzing and interpreting data
- Constructing explanations
- Engaging in argument from evidence
- Obtaining, evaluating, and communicating information

Disciplinary Core Ideas

LS1: How do organisms live, grow, respond to their environment, and reproduce?
LS1.A: Structure and function
LS1.C: Organization for matter and energy flow in organisms
ESS2: How and why is Earth constantly changing?
ESS2.E: Biogeology
ESS3: Earth and human activity
ESS3.A: Natural resources

Crosscutting Concepts

- Patterns
- Cause and effect
- Systems and system models
- Structure and function

FOSS Full Option Science System

INVESTIGATION 2 – Water and Land Snails

	Investigation Summary	Time	Focus Question for Phenomenon, Practices
PART 1	**Observing Water Snails** Students are introduced to two kinds of aquatic snails. They investigate the snails' physical characteristics and behavior for similarities and differences. They find out about the snails' needs to live and grow.	**Introduction** 5 minutes **Center** 15–20 minutes **Notebook** 15 minutes	**What are the parts of a water snail?** **Practices** Asking questions Planning and carrying out investigations Analyzing and interpreting data
PART 2	**Shells** Students observe seashells. Using their experience with living snails, they look for shells that they think might have belonged to relatives of the water snail they observed. They organize the shells into pairs or groups and give rationales for their decisions.	**Introduction** 10 minutes **Center** 15–20 minutes **Notebook** 15 minutes	**How can shells be grouped?** **Practices** Planning and carrying out investigations Analyzing and interpreting data
PART 3	**Land Snails** Students collect and get to know local land snails. They handle the snails, observe their features, and see how they interact with objects. They compare their structures and behaviors to the water snails. They compare what the land snail needs to live and grow to the water snail.	**Introduction** 5 minutes **Outdoors** 15 minutes **Center** 12–15 minutes **Notebook** 15 minutes **Reading** 15 minutes	**What do land snails do?** **Practices** Asking questions Planning and carrying out investigations Analyzing and interpreting data Constructing explanations Engaging in argument from evidence Obtaining, evaluating, and communicating information

At a Glance

Content Related to DCIs	Writing/Reading	Assessment
• Snails are animals and have basic needs—water, air, food, and space with shelter. • Different kinds of snails have some structures and behaviors that are the same and some that are different.	**Science Notebook Entry** *Water-Snail Outline*	**Embedded Assessment** Teacher observation
• There is great diversity among snails. • Shells differ in size, shape, pattern, and texture.	**Science Notebook Entry** Draw or write words to answer the focus question. **Video** *Seashore Surprises*	**Embedded Assessment** Teacher observation
• Different kinds of snails have some structures and behaviors that are the same and some that are different. • Snails have senses. • Snails are animals and have basic needs—water, air, food, and space with shelter.	**Science Notebook Entry** Draw or write words to answer the focus question. **Science Resources Book** "Water and Land Snails"	**Embedded Assessment** Teacher observation **NGSS Performance Expectations addressed in this investigation** K-LS1-1 K-ESS2-2 K-ESS3-1

Animals Two by Two Module—FOSS Next Generation

INVESTIGATION 2 – Water and Land Snails

BACKGROUND for the Teacher

Snails are not everybody's favorite animals. They are not what you would call lively, and they don't do tricks. They are not particularly popular with gardeners because they eat many types of plants. To make matters worse, snails leave their signature wherever they travel—the familiar mucous trail. However, early-childhood students will find them interesting and irresistible, even if some adults struggle with the snail's bad image. And they exhibit a shell—a phenomenon students will investigate.

What Are the Parts of a Water Snail?

Snails are mollusks, along with clams, oysters, slugs, squids, and octopuses. Snails and slugs are in a class called gastropods, meaning "belly-foot." The muscular part of the snail on which it typically slides is the **foot**, and because it is where any upright animal might have its belly, it is easy to see why the name arose. Many gastropods have no **shell**, but the vast majority of them carry their home on their back, and usually the shell is coiled up. Coiling is a very efficient way to engineer an ever-expanding home, because the animal uses the outside of the previous turn of shell as one wall of the new extension. As long as the snail continues to grow, it will continue to lay new shell material around the opening. The shell is always **large** enough to allow the snail to withdraw inside when a predator threatens or when the environment is harmful.

Garden and woodland snails are definitely in the gastropod minority. Most of the class are at home in the ocean, where snails evolved. Those that crept out on the land developed lungs. In time some of those invaded fresh water but retained a kind of lung, so they must rise to the surface to breathe air with some frequency. Many freshwater snails and all marine snails, however, breathe with gills, so they are truly aquatic animals.

The pond snails and ramshorn snails students will observe in this module do not breathe using gills. Their gills are reduced, and the snail has only a thin wall of the mantle cavity through which to assimilate oxygen. This organ is called the snail's lung, and this type of snail belongs to the pulmonate (lung) snails. The genus is *Pulmonata*. The advantage to breathing with lungs is that the snails can breathe oxygen from dry air and do not have to rely on water. To do this, freshwater snails have to regularly go to the water surface. They climb plants or just **float** to the surface of the water to get oxygen. Snails make use of water tension at the surface to hang **upside down**, sometimes eating algae and getting oxygen at the same time. Some gill-breathing snails exhibit this behavior as well but use it more to get food than to get oxygen.

"*Snails and slugs are in a class called gastropods, meaning 'belly-foot.'*"

Background for the Teacher

Ramshorn and pond snails have **tentacles**, two of them, that cannot be withdrawn. Each tentacle has an eye at the base. **Water snails** glide along on a mucous trail like their terrestrial cousins, but it is hard to see the trail in the water. Occasionally an aquatic snail will travel up and down mysteriously through open water. In actuality the snail is gliding on invisible strands of mucus.

Most aquatic snails lay eggs; they are smaller than land-snail eggs and are deposited on the surfaces of plants, rocks, or aquarium walls in a thick gelatinous mass. Again, the snails hatch out as tiny editions of the adults, to go about their business of eating decomposing plant and animal material and scraping algae from solid surfaces with their radula.

Water snails are very easy to care for. All you need is a container of aged or chemically dechlorinated (conditioned) water. Keep plenty of aquatic plants in the aquarium for them to eat and to lay their eggs on. Water snails will also eat lettuce, spinach, and zucchini, or fish-feeding cakes and tiny bits of flake fish food you can buy at the aquarium store or pet store. Although water snails eat waste materials and algae, thus helping to keep the aquarium looking clean, it is best to add fresh water once a week to keep them healthy. Be sure to use conditioned water.

When the investigations are completed, you can keep the snails in an aquarium with fish or in their own aquarium. A jar or other clear container will work fine. They require very little attention—just maintain the water level and add a bit of food now and again. With any luck at all, you will have an ample permanent supply of snails for your use and enough to share with the other teachers at your school.

How Can Shells Be Grouped?

Sea shells come in many sizes, shapes, colors, and textures, and these properties can be used to sort shells. Even within one kind of shell, there will be variation in size and color. Students will find many ways to group a collection of shells.

Animals Two by Two Module—FOSS Next Generation

INVESTIGATION 2 – *Water and Land Snails*

What Do Land Snails Do?

Land snails have a head with a mouth and two pairs of tentacles. The larger pair is higher on the head, and each is tipped with a primitive eye. The eye cannot render an image for the snail, but it is sensitive to **light**. The smaller pair of tentacles is directed earthward and serves to sense food and terrain. When the snail is threatened, it can withdraw the tentacles into the tissue of its head, and if it is further harassed, it can pull its head and foot completely into its shell.

We all know about a snail's pace—pretty slow. A typical snail going at a good clip will leave a trail about 27 kilometers (km) long in a year. That is slow. If you have a snail in a cup and you leave the room for a few minutes, however, you will be amazed at how far it can go. During the course of one night, the time that snails are most active, they can cover enough ground to encounter and consume all of the lettuce seedlings in your garden.

Lettuce is one of the snail's favorite foods and a good food source for captive snails in the classroom, but they will eat a wide variety of plants. A snail eats by pressing its mouth against the intended meal and scraping the food material to bits with its specialized rasplike tongue, called a radula. As the sun approaches, the sated snail heads for cover under a ledge or behind some leaves. There it will settle down, seal itself with mucus to some surface, and wait for sunset.

If the weather is cold or dry, two conditions that are unfavorable for snail activities, the snail might lapse into a deep resting state called estivation. The condition is similar to hibernation, but the factors that induce it are different. Snails estivate to conserve moisture and close out harmful predators, parasites, and diseases.

Most land snails are hermaphroditic, meaning that every snail has both eggs and sperm. Snails do have to mate, however, because it is not possible for a snail to fertilize its own eggs. After mating, the snail lays up to 100 tiny round eggs on or just under the surface of the soil. Two to four weeks later they hatch as tiny replicas of the adult, frail, vulnerable, and hungry. Soon the shell hardens, and away they slide to find a head of lettuce.

Background for the Teacher

Keep land snails in a clear basin with damp paper towels in the bottom and a lid securely fixed over the top. Snails are very good at climbing out of containers! The lids supplied in the kit have plenty of holes so the snails will get air. Check the paper towels daily and make sure they remain damp. You can lay a sheet of plastic wrap loosely over the top of the lid to maintain the moisture level.

Land snails eat green leafy plants. The **dark**-green outer leaves of lettuce that most grocery stores throw away are ideal for feeding snails. Ask the produce person for the outer leaves. Occasionally put some eggshell or clean white chalk in the **terrarium** for the snails to eat as a source of calcium for strong shells.

If the snails are estivating (sealed up and inactive) at the time you want to use them, simply dunk them in a cup of tap water for 5 seconds and return them to the terrarium. They will begin to move around in a minute or so.

Put the snails in a soil-filled terrarium after students have finished their investigations if you want to see the reproductive process. When snails lay eggs, they generally bury them in soil or lay them under leaves or pieces of wood; you might not see them. When the tiny snails emerge, move them to a separate terrarium so the larger snails don't crush them. They survive under the same conditions as the adult snails.

There are a number of options for dealing with the land snails at the end of the module. The organisms might find a permanent home in your classroom. That would be ideal, as you can continue informal observations for a longer time. You will need to provide containers for permanent habitats if the kit will be used by another teacher. Or you might pass the organisms on to the next user of the kit. Or you can check with your district to see if there is a plan for reuse of FOSS organisms.

If your students collected the snail from the local environment, it is OK to return them to that environment. If you did not collect the snails locally, do NOT release them into the environment. Snails should never be released where they were not collected.

If there is no other option for the organisms, euthanize them by placing them in the freezer for a day or two, and then dispose of them in the trash.

"Put the snails in a soil-filled terrarium after students have finished their investigations if you want to see the reproductive process."

Animals Two by Two Module—FOSS Next Generation

INVESTIGATION 2 – *Water and Land Snails*

TEACHING CHILDREN *about* Snails

Developing Disciplinary Core Ideas (DCI)

In this module we strive to provide young students with opportunities to experience several kinds of animals up close. This includes handling the animals carefully whenever appropriate. Some students will approach animals like snails enthusiastically, picking them up right away and eagerly letting them move on their hands. Others will have very different responses. Some will touch but will be reluctant to pick up the animals, and others will not want to touch them because they are afraid or perceive them to be disgusting.

Don't force students to pick up the snails. Reluctant touchers can observe snails in a clear cup or on a piece of paper, some cardboard, or a stick. Students should be encouraged to work with the animals in ways that are comfortable for them. In many cases, students will gain courage as they observe their peers handling the snails and will eventually join in the fun.

Students are sometimes put off by the slime trail produced by land snails. They might think it makes the snail dirty, "gross," or "yucky." This is the time for you to provide information to the contrary, telling students that the slime trail is the snail's path (made mostly of water) and the path is necessary for the snail to slide smoothly along many surfaces. Reinforce that the trail is clean and can be washed off any surface with water.

As usual, one of the underlying goals of any activity where early-childhood students work with animals is to build respect for life. Living organisms constitute the biological wealth of the planet, and early exposure to attitudes that respect and nurture all life-forms will contribute to a population able to manage that wealth sensitively. Students might be confused if they come from homes where snails are looked upon as pests to be eliminated. Help students understand that in the classroom, the snails are respected and honored as living organisms that have needs. It is the responsibility of students to care for the animals so they will prosper.

The activities and readings students experience in this investigation contribute to the disciplinary core ideas **LS1.A, Structure and function: All organisms have external parts; LS1.C, Organization for matter and energy flow in organisms: All animals need food in order to grow; ESS2.E, Biogeology: Plants and animals can change their environment;** and **ESS3.A, Natural resources: Living things need water, air, and resources from the land, and they live in places that have the things they need.**

NGSS Foundation Box for DCI

LS1.A: Structure and function
- All organisms have external parts. Different animals use their body parts in different ways to see, hear, grasp objects, protect themselves, move from place to place, and seek, find, and take in food, water, and air. Plants also have different parts (roots, stems, leaves, flowers, fruits) that help them survive and grow. (foundational)

LS1.C: Organization for matter and energy flow in organisms
- All animals need food in order to grow. They obtain their food from plants or from other animals. Plants need water and light to live and grow. (K-LS1-1)

ESS2.E: Biogeology
- Plants and animals can change their environment. (K-ESS2-2)

ESS3.A: Natural resources
- Living things need water, air, and resources from the land, and they live in places that have the things they need. Humans use natural resources for everything they do. (K-ESS3-1)

Teaching Children about Snails

Engaging in Science and Engineering Practices (SEP)

In this investigation, students engage in these practices.

- **Asking questions** about snail structures, behaviors, what they eat, and where they live.
- **Planning and carrying out investigations** with water snails and land snails to observe their structures and study their needs.
- **Analyzing and interpreting data** by describing observations of snails over time, recording information, using and sharing notebook entries, including writing and labeled pictures. Students use their firsthand observations and those of others in the classroom to describe the patterns they observe in snail aquaria.
- **Constructing explanations** by making firsthand observations of water snails and land snails and using this as evidence to answer questions about the needs of animals, including food.
- **Engaging in argument from evidence** to support a claim by discussing preferences of land snails based on firsthand observations.
- **Obtaining, evaluating, and communicating information** about structures of water snails and land snails, their needs, and where they live.

NGSS Foundation Box for SEP

- **Ask questions** based on observations to find more information about the natural and/or designed world(s).
- **With guidance, plan and conduct an investigation** in collaboration with peers (for Grade K).
- **Make observations** (firsthand or from media) and/or measurements to collect data that can be used to make comparisons.
- **Make predictions** based on prior experiences.
- **Record information** (observations, thoughts, and ideas).
- **Use and share pictures, drawings,** and/or writings of observations.
- **Use observations (firsthand or from media)** to describe patterns in the natural world in order to answer scientific questions.
- **Compare predictions** (based on prior experiences) to what occurred (observable events).
- **Make observations** (firsthand or from media) to construct an evidence-based account for natural phenomena.
- **Construct an argument** with evidence to support a claim.
- **Read grade-appropriate text** and/or use media to obtain scientific and/or technical information to describe patterns in the natural world.
- **Communicate** information or solutions with others in oral and/or written forms using models and/or drawings that provide detail about scientific ideas.

INVESTIGATION 2 – Water and Land Snails

NGSS Foundation Box for CC

- **Patterns:** Patterns in the natural and human-designed world can be observed, used to describe phenomena, and used as evidence.
- **Cause and effect:** Events have causes that generate observable patterns. Simple text can be designed to gather evidence to support or refute student ideas about causes.
- **Systems and system models:** Objects and organisms can be described in terms of their parts. Systems in the natural and designed world have parts that work together.
- **Structure and function:** The shape and stability of structures of natural and designed objects are related to their function(s).

Exposing Crosscutting Concepts (CC)

In this investigation, the focus is on these crosscutting concepts.

- **Patterns.** Structures of snails are similar but they have differences in how they look, the appearance of their shells (size, shape, color, pattern), and where they live.
- **Cause and effect.** Snails can change their water environment over time.
- **Systems and system models.** Snails can be described in terms of their structures.
- **Structure and function.** The observable structures of snails (shell, foot, antennae, head) serve functions in survival.

Connections to the Nature of Science

This investigation provides connections to the nature of science.

- **Scientific investigations use a variety of methods.** Scientific investigations begin with a question. Scientists use different ways to study the world.
- **Scientific knowledge is based on empirical evidence.** Scientists look for patterns and order when making observations about the natural world.

New Word (Say it, See it, Hear it, Write it)

Dark
Float
Foot
Land snail
Large
Light
Rough
Sea animal
Shell
Sideways
Small
Smooth
Snail
Tentacle
Terrarium
Upside down
Vial
Water snail

Teaching Children about Snails

Conceptual Flow

Students continue to explore how animals of all kinds have various needs to live and grow. To understand an animal's needs, you need to first get to know the animal—its structures and behaviors. The second group of animals students study and care for have members that live primarily in the water and members that live primarily on land. Snails are the main phenomenon in this experience. The guiding question is what do animals such as snails need to live and grow?

The **conceptual flow** for this investigation starts with an introduction to **water snails**. Students observe several water snails of different kinds, describe their parts (structures), and compare the variations in size, color, and shape. Students learn how to care for the snails and set up an aquarium to meet the snails' needs.

In Part 2, students are provided with a collection of sea shells. Students observe the **shell properties** and group them into kinds.

In Part 3, students go outdoors to find and collect local land snails. Students compare the structures and behaviors of the land snails to the water snails and find out what the land snails need to live and grow.

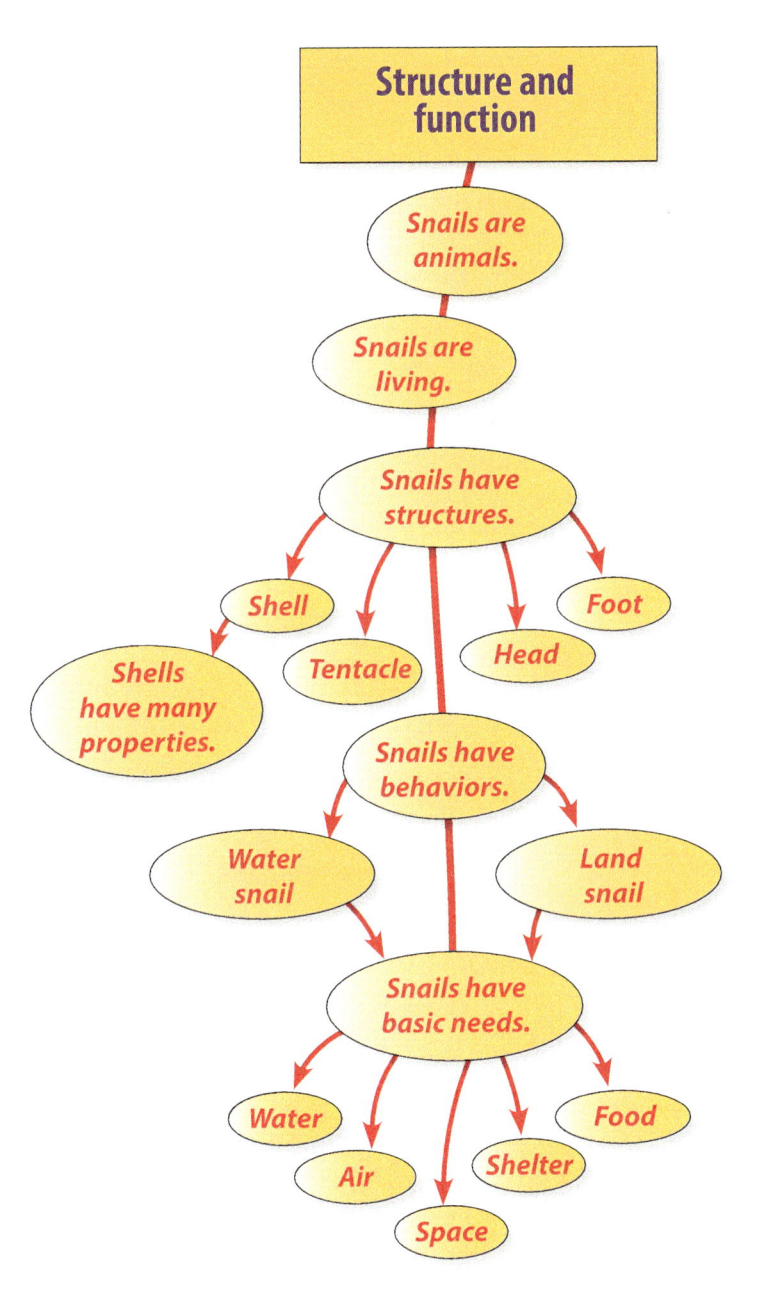

Structure and function
- Snails are animals.
- Snails are living.
- Snails have structures.
  - Shell
    - Shells have many properties.
  - Tentacle
  - Head
  - Foot
- Snails have behaviors.
  - Water snail
  - Land snail
- Snails have basic needs.
  - Water
  - Air
  - Space
  - Shelter
  - Food

Animals Two by Two Module—FOSS Next Generation

INVESTIGATION 2 – Water and Land Snails

MATERIALS for
Part 1: Observing Water Snails

For each student at the center
- 1 Vial, 7 dram
- 1 *Water-Snail Outline* ★

For the class
- 1 Goldfish or guppy aquarium (from Investigation 1)
- 1 Clear basin and cover
- 1 Plastic spoon
- 12 Ramshorn snails (See Step 4 of Getting Ready.) ★
- 12 Pond snails (See Step 4 of Getting Ready.) ★
- 1 Bunch of elodea ★
- 1 Bottle of dechlorinator
- • Gravel
- 1 Tunnel
- • Paper towels ★
- • Aged water ★
- 5 Large bug boxes
- ❏ 1 Teacher master 12, *Center Instructions—Observing Water Snails*
- ❏ 1 Teacher master 13, *Water-Snail Outline*

For assessment
- ❏ • *Assessment Checklists* 1 and 2

★ Supplied by the teacher.   ❏ Use the duplication master to make copies.

No. 12—Teacher Master

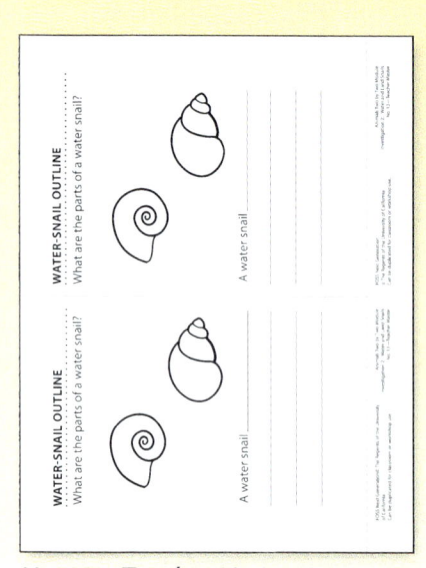

No. 13—Teacher Master

Part 1: Observing Water Snails

GETTING READY for
Part 1: Observing Water Snails

1. **Schedule the investigation**
   Each group of six to ten students will need 15–20 minutes with the water snails. Plan an additional 5 minutes with the whole class to introduce the center and 15 minutes for students to write or draw in their notebooks.

2. **Preview Part 1**
   Students are introduced to two kinds of aquatic snails. They investigate their physical characteristics and behavior for similarities and differences. The focus question is **What are the parts of a water snail?**

3. **Set up the aquarium**
   Use a clear basin to set up one aquarium for pond snails. Fill the container with about 5 liters (L) of aged tap water, or use the dechlorinator provided in the kit to condition the water. Put rinsed gravel and a tunnel in the aquarium. Add some elodea as well.

4. **Obtain water snails**
   You will need to obtain two types of aquatic snails for this part. Ramshorn snails (*Planorbis rubrum*) and pond snails (*Lymnaeidae*) can be ordered from Delta Education. Approximately 12 dime-sized snails are in a set. You will need one set of each type. The ramshorn snail has the familiar spiral-horn-shaped shell. The pond snail is more dome-shaped but is also spiraled toward the tip.

   If you bought elodea for your fish at an aquarium or pet store, you might already have a few water snails! All snails can go into the prepared aquarium.

5. **Set up vials with snails**
   Just before the investigation, prepare a set of vials or bug boxes with a sprig of elodea and aged tap water (you can use the water from the snail aquarium) so that each student at the center will have one to observe. Get three ramshorn and three pond snails. Place one snail in each vial. Make sure there is a layer of air at the top of each vial and place the caps on. Remember, the snails breathe air.

▶ **NOTE**
This is a good time to move the surviving goldfish from two aquariums into one aquarium. This will free up an aquarium for water snails.

▶ **NOTE**
To prepare for this investigation, view the teacher preparation video on FOSSweb.

Ramshorn snail

Pond snail

▶ **NOTE**
When all students have completed this activity, transfer some of the water snails to the goldfish and guppy aquariums.

Animals Two by Two Module—FOSS Next Generation

INVESTIGATION 2 – Water and Land Snails

6. **Plan assessment for Part 1**

    Plan to listen and observe students as they work at the center and to review their notebook entries during or after class. Record your observations on *Assessment Checklists* 1 and 2.

    ### What to Look For

    - *Students ask questions about water snail structures. (Asking questions; structure and function.)*

    - *Students describe observations of the structures of water snails and record information by writing and labeling pictures. (Planning and carrying out investigations; LS1.A: Structure and function; patterns.)*

Part 1: Observing Water Snails

GUIDING *the Investigation*
Part 1: *Observing Water Snails*

1. **Introduce the water snails**
   Call students to the rug. Tell them that you have brought a new animal to class for them to observe using their senses. Show the guppy or goldfish aquarium and tell them this animal lives in water with fish.

   Tell them,

   This animal has no arms or legs. It moves slowly. It carries its house on its back.

   ➤ *What kind of animal is it?* [Water snail.]

   Bring the water-snail aquarium to the rug. Explain that over the next few weeks, they will be finding out what snails need to live and grow. Tell students that their task at the center today will be to get to know the snails by observing them closely and finding out what body parts, or structures, snails have and what snails do. Explain that because these snails live in water like fish, it is best not to touch them. Students will be able to observe them closely in the aquariums, in vials, and in special boxes with magnifiers.

   Review proper handling of living things to ensure that the snails will not get injured.

2. **Move to the center**
   Send one group to the center. Let students observe the water snails in the aquarium for a few minutes without a lot of guidance. Don't tell students that there are two different types of water snails. Let them discover it for themselves. Have them share their observations. If students don't mention it, have them notice the location of the snails in the aquarium.

3. **Observe snails in vials**
   Give each student at the center one small vial (or large bug box) with a snail and elodea. Half the students will have one type of snail, and half the other type of snail. Guide their observations.

   ➤ *Where are the **snails** in the **vials**? Are they **floating** in the water or holding onto the sides?*

   ➤ *Do you see a **shell**? A **foot**?*

   ➤ *Does the snail have **tentacles** (feelers)? Where are they?*

   ➤ *What else do you observe?*

   ➤ *Does the snail have a head? A tail? Eyes? A mouth?*

   Animals Two by Two Module—FOSS Next Generation

FOCUS QUESTION
What are the parts of a water snail?

EL NOTE
Use gestures to show the structures and behaviors of the water snail.

SCIENCE AND ENGINEERING PRACTICES
Planning and carrying out investigations

Materials for Steps 1 and 3
- *Water-snail aquarium*
- *Fish aquarium*
- *Large bug boxes with elodea and snails*
- *Vials with elodea and snails*
- *Paper towels*

Say it / See it / Hear it / Write it
New Word

133

INVESTIGATION 2 – Water and Land Snails

> Is the snail moving? How does it move?

Encourage students to ask questions of their own.

4. **Compare two types of water snails**
   Have each student trade snails with another student so they can observe the other type. Ask questions to guide their comparison.

   > Does this snail look the same as the other one?

   > How is it different?

   > What else do you observe?

   > Does this snail have a shell? A foot? Tentacles?

   > What questions do you have about snails?

5. **Put snails back into the aquarium**
   After students have observed both kinds of water snails, ask,

   > What do you think will happen if we put two of the snails back in the aquarium?

   Listen to students' ideas and then tell them you are going to carefully drop two snails into the water. If you can, drop them on top of the tunnel or in a corner as a reference point.

   Students might be surprised to see that the snails have stuck to the side of the vial when you attempt to pour them into the tank. Ask students what made the snails stick to the sides of the vials. Use the end of a plastic spoon to gently remove a snail from the vial.

6. **Focus question: What are the parts of a water snail?**
   Write the focus question on the chart as you read it aloud.

   > What are the parts of a **water snail**?

   Distribute teacher sheet 13, *Water-Snail Outline*, to each student. Describe how to glue the sheet into the notebook before they answer the question (the question is on the sheet). When they return to their tables, they record their observations in their notebooks with pictures and words. You might want to model a response.

7. **Prepare for the next group**
   Have students take one last look at the aquarium and return the vials with snails to the center of the table. The next group can come up to the center. You might need to refill the two vials with snails and elodea if you don't have enough for the new group to observe.

TEACHING NOTE

It is OK for students to place the water snails on their hands or on the table for short observations.

SCIENCE AND ENGINEERING PRACTICES

Asking questions

Planning and carrying out investigations

Analyzing and interpreting data

Materials for Step 5
- Plastic spoon

New Word — Say it, See it, Hear it, Write it

CROSSCUTTING CONCEPTS

Systems and system models

Structure and function

Full Option Science System

Part 1: Observing Water Snails

8. **Clean up**
   When the last group has completed their observations, have them carefully return the snails and plants to the aquariums. You might need to assist them if any snails are stuck to the sides of the vials. Have students return the empty vials and caps to a designated spot at the center. A few snails can be added to the goldfish and guppy aquariums.

9. **Review vocabulary**
   Gather students at the rug. Review key vocabulary added to the word wall. Here's a suggested cloze review. Students answer chorally.

   ➤ *Our new animal is a _____.*

   S: Snail.

   ➤ *These snails live in _____.*

   S: Water.

   ➤ *The large surface on which the snail moves is called its _____.*

   S: Foot.

   ➤ *The protective covering on a snail is called a _____.*

   S: Shell.

   ➤ *Extending from its head, a water snail has _____. They are used for sensing its surroundings.*

   S: Tentacles.

10. **Compare the two water snails**
    To summarize the comparisons that students made, draw a T-table to list differences, as you did with the fish comparison. Ask students to offer ideas for the "differences" chart.

    ➤ *How are the two water snails different?*

    On a separate sheet, list the similarities based on students sharing their observations.

    ➤ *How are the two water snails the same?*

11. **Have a sense-making discussion**
    Model how to add one or two of the parts from the table to the class notebook to add to the focus question answer.

    Ask students to select one of the parts that are the same for the snails, such as shell, foot, head, tentacles, and to talk with a partner about what that part helps the snail to do.

EL NOTE

Write the cloze sentences on chart paper with an illustration. Have the vocabulary words on cards or sentence strips as a reference.

- float
- foot
- shell
- snail
- tentacle
- vial
- water snail

CROSSCUTTING CONCEPTS

Patterns

Ramshorn snail	Pond snail
Dark brown shell	Olive green shell
Spiral shell	Domed shell
Moves all over aquarium	Moves on bottom or sides of aquarium
Narrow foot	Broad, oval foot
Tail and head beyond shell when moving	Tail and head close to shell

SCIENCE AND ENGINEERING PRACTICES

Analyzing and interpreting data

TEACHING NOTE

Refer to the Sense-Making Discussions for Three-Dimensional Learning chapter in Teacher Resources on FOSSweb for more information about how to facilitate this with young students.

Animals Two by Two Module—FOSS Next Generation

INVESTIGATION 2 – Water and Land Snails

12. Discuss snail needs
Ask students,

➤ *What do water snails need to live?*

Give students time to share their ideas with a partner. Students should build on what they know about goldfish and guppies to say these snails need water (with oxygen), food (plants or fish food), space, and shelter. Ask students to list the parts of the aquarium to see if the snails have what they need. Ask students if they have any questions about the snail structures, behaviors, and how it lives.

WRAP-UP/WARM-UP

13. Share notebook entries
Conclude Part 1 or start Part 2 by having students share notebook entries. Ask students to open their science notebooks to the last entry. Read the focus question together.

➤ *What are the parts of a water snail?*

Ask students to pair up with a partner to

- share their answers to the focus question;
- explain their drawings.

Encourage students to revise their entries by labeling the snail parts using the vocabulary words from the word chart, adding where snails live (water, elodea). Ask students to discuss what they think the parts of the water snail are used for.

ELA CONNECTION

This suggested strategy addresses the Common Core State Standards for ELA.

SL 1: Participate in collaborative conversations.

W 5: Strengthen writing.

Part 2: Shells

MATERIALS for
Part 2: *Shells*

For each student at the center
1 Sheet of construction paper ★

For the class
1 Bag of assorted shells
1 Computer with Internet connection ★
1 Projection system ★
❑ 1 Teacher master 14, *Center Instructions—Shells*

For assessment
- *Assessment Checklists* 1 and 2

★ Supplied by the teacher.   ❑ Use the duplication master to make copies.

No. 14—Teacher Master

Animals Two by Two Module—FOSS Next Generation

INVESTIGATION 2 – *Water and Land Snails*

GETTING READY *for*
Part 2: *Shells*

1. **Schedule the investigation**
   Each group of six to ten students will need 15–20 minutes at the center. Plan an additional 10 minutes with the whole class to introduce the center and 15 minutes for students to write or draw in their notebooks.

2. **Preview Part 2**
   Students observe seashells. Using their experience with living snails, they look for shells that they think might have belonged to relatives of the water snail they observed. They organize the shells into pairs or groups and give rationales for their decisions. The focus question is **How can shells be grouped?**

3. **Prepare for the introduction**
   Select about eight shells from the collection to show students when you introduce the activity. Choose three that are from the same kind of snail, three random kinds of snails, and two or three from other mollusks (flat shells instead of spiraled).

4. **Plan for viewing video**
   Preview the video *Seashore Surprises*, chapters 1–2, Characteristics of a Seashore and Learning about Seashells (duration 8 minutes). Prepare to use that to introduce the seashells to the whole class before students explore the shells. The link to this video in the Resources by Investigation section of FOSSweb.

5. **Plan assessment for Part 2**
   Plan to listen and observe students as they work at the center and to review their notebook entries during or after class. Record your observations on *Assessment Checklists* 1 and 2.

   What to Look For

   - *Students analyze and organize snail shells into groups or seriate them by a single property and explain their reasoning. (Analyzing and interpreting data; LS1.A: Structure and function; patterns.)*

Part 2: Shells

GUIDING *the Investigation*
Part 2: *Shells*

1. Introduce the video
Call students to the rug. Invite them to come with you on a trip to the seashore. Show them chapters 1 and 2 of *Seashore Surprises*.

Chapter 1: Characteristics of a Seashore (3 min 13 sec)
- A seashore is defined as the area where the land and sea meet; seashores have unique ecosystems; habitats.
- Although sea coral looks like a tree branch, it is actually made from tiny sea animals.
- Ocean waves constantly transport and deposit seashells along the shoreline.

Chapter 2: Learning about Seashells (5 min 11 sec)
- Seashells are made by sea animals that grow the shells for protection.
- Many seashells found along the beach are empty, but the host finds a banded tulip; he shows how this shellfish uses its trapdoor for protection from predators.
- An olive shell is very smooth and shiny.
- The spiny oyster has spines sticking out from his shell, which help to protect it from predators.
- The animal in the pen shell coats the inside of its shell with a liquid; when it hardens, it is known as mother of pearl; kitten's paw shells got their name because they look like a kitten's paw.
- Coquinas are tiny shellfish that use a foot to burrow into the sand and escape from their predators.
- Lightning whelks get larger as they grow; the whelk lays hundreds of eggs in a special protective casing.

FOCUS QUESTION
How can shells be grouped?

Animals Two by Two Module—FOSS Next Generation

INVESTIGATION 2 – Water and Land Snails

Materials for Step 2
- *Construction paper*
- *Shells*

> **NOTE**
> Each student should get about 20 shells; there are about 400 in the kit.

EL NOTE
Allow students more time to observe and describe what they notice. Encourage them to ask questions about the shells.

SCIENCE AND ENGINEERING PRACTICES
Planning and carrying out investigations

CROSSCUTTING CONCEPTS
Patterns

2. **Introduce shells**
   Show students the eight shells that you selected. Have students discuss with a partner if they know where the shells might have come from and then share with the class. [The ocean.] Ask,

   ➤ *Do you think that all of these shells came from the same kind of sea animal? Why or why not?* [The spiral shells are from different kinds of snails. Some might be from the same kind of snail but are a little different from one another; no two shells are *exactly* alike. The flat shells are from other kinds of sea animals.]

   Tell students that their task will be to carefully observe the shells using their senses and to group them in some way. How students put them together will be up to them.

3. **Move to the center**
   Direct a small group of six to ten students to the center. Make sure each is seated within reach of the shells, and that each has a sheet of construction paper to work on.

4. **Distribute shells**
   Tell students that there are plenty of shells for everyone and that they should be handled gently so they won't break. If students are concerned about uneven distribution, ask them to think of a way to divide the shells so that everyone gets 20 shells.

5. **Observe shells**
   When everyone has his or her set of shells, have students observe the shells closely. Distribute a piece of construction paper and ask students how they could organize their shells. If necessary, ask these questions.

   ➤ *How many shells do you have?*
   ➤ *Can you find two that look alike?*
   ➤ *Can you find two shells that came from different kinds of snails?*
   ➤ *Are any of the shells like the water snails you observed?*
   ➤ *Can you find any shells that you think came from sea animals other than snails?*

6. **Discuss shell organization**
   Ask students to arrange their shells in some way. They may choose to put them together in pairs, make some sort of design, or form a pattern using the shells. As each student completes this task, ask him or her to explain the arrangement.

Part 2: Shells

7. **Seriate shells**
   Ask students to put the shells in order from the **largest** to the **smallest**, from **darkest** to **lightest**, **rough** to **smooth**, or any other way they like.

8. **Review vocabulary**
   Review key vocabulary added to the word wall.

9. **Focus question: How can shells be grouped?**
   Write the focus question on the chart as you read it aloud.

   ➤ How can shells be grouped?

   Tell students that you have a strip of paper with the focus question written on it. Describe how to glue the strip into the notebook before they answer the question. When they return to their tables, they should answer the focus question in their notebooks with pictures and words. You might want to model a response by making an outline of a shell and providing a sentence frame such as, Shells can be grouped by _____ . or My drawing shows _____ .

10. **Prepare the center for the next group**
    Have students return all the shells to the center of the table. They can leave their construction paper where it is.

WRAP-UP/WARM-UP

11. **Share notebook entries**
    Conclude Part 2 or start Part 3 by having students share notebook entries. Ask students to open their science notebooks to the last entry. Read the focus question together.

    ➤ How can shells be grouped?

    Ask students to pair up with a partner to
    - share their answers to the focus question;
    - explain their drawing;
    - share one new thing they learned about shells.

SCIENCE AND ENGINEERING PRACTICES
Analyzing and interpreting data

dark
large
light
rough
sea animal
small
smooth

FOCUS CHART

How can shells be grouped?

By size, shape, color, pattern.

Animals Two by Two Module—FOSS Next Generation

INVESTIGATION 2 – *Water and Land Snails*

MATERIALS for
Part 3: *Land Snails*

For each pair of students

- 1 Plastic cup with lid
- 1 Piece of cardboard, 10 × 30 cm (4" × 12")
- 2 *Land-Snail Outline* ★
- 2 *FOSS Science Resources: Animals Two by Two*
  - "Water and Land Snails"

For the class

- 12–15 Local land snails (See Step 4 of Getting Ready.) ★
- 1 Clear basin with cover
- • Plastic wrap
- 2 Containers with lids, 1/2 L
- 1 Spray mister
- • Chalk or egg shells ★
- • Paper towels ★
- • Lettuce or carrot ★
- ❏ 1 Teacher master 15, *Center Instructions—Land Snails*
- ❏ 1 Teacher master 16, *Land-Snail Outline*
- 1 Big book, *FOSS Science Resources: Animals Two by Two*

For assessment

- *Assessment Checklists* 1, 2, and 3

★ Supplied by the teacher. ❏ Use the duplication master to make copies.

No. 15—Teacher Master

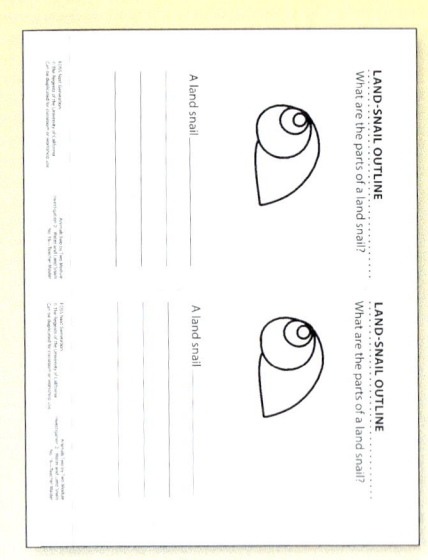

No. 16—Teacher Master

142 Full Option Science System

Part 3: Land Snails

GETTING READY *for*
Part 3: *Land Snails*

1. **Schedule the investigation**

   This part starts outdoors to collect schoolyard land snails for observation in the classroom. Plan for 15 minutes for the search. You can conduct the land-snail search and/or the observation sessions with the whole class or with smaller groups using a center. Each group of six to ten students will need 12–15 minutes to observe snails if you use a center. Plan an additional 5 minutes with the whole class to introduce the center and 15 minutes for students to write or draw in their notebooks. Plan another session of 15 minutes for the reading.

2. **Preview Part 3**

   Students collect and get to know local land snails. They handle the snails, observe their features, and see how they interact with objects. They compare their structures and behaviors to the water snails. The focus question is **What do land snails do?**

3. **Select your outdoor site**

   Check around the school for land snails. Search bushes, hedges, and along the bottoms of fences in moist areas and in crevices along edges of garden beds. If you have a school garden, look there first. Snails will hide in damp leaves and at the base of plants. If you can find 12–15 snails that are big enough for students to handle, plan to take the class out to collect snails.

   Determine the outdoor boundaries for the activity and plan to describe them so students know where to search for snails.

4. **Obtain land snails**

   When you find land snails in your schoolyard, collect one or two large ones and place them in the terrarium (see Step 5) for the introduction in Step 1 of Guiding the Investigation.

   If you can't find snails in your schoolyard, check your own garden or that of a friend or colleague, or a local park. Older students can be hired to round up a few snails for you.

   If no snails are to be found locally, then consider ordering them from a biological supplier within your state. Because land snails can become garden pests, the movement of snails across state boundaries is carefully controlled, and permits are needed to order them from outside your state. The best way to get land snails into your classroom is to collect them locally.

   If you don't have snails but have slugs, use them instead in this investigation.

▶ **NOTE**
You can provide snails for students to observe in class or have students collect them from the schoolyard. See the Plant and Animal care pages on FOSSweb for more information on obtaining land snails.

Animals Two by Two Module—FOSS Next Generation

INVESTIGATION 2 – Water and Land Snails

> **NOTE**
> Some helpers might be reluctant to touch animals. Remind them that their job is to keep students safe, and that they don't have to touch the snails. They are there to support students' enthusiasm for the snail search.

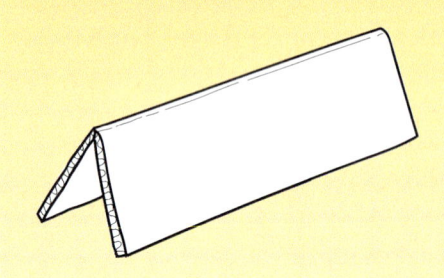

5. **Plan for the land snail terrarium**
   Put two or three damp paper towels in the bottom of a clear basin. Plan to feed the snails bits of lettuce or carrot. Be diligent about keeping the lid on the basin; otherwise your activity might start with a snail hunt. Lay a sheet of plastic wrap loosely over the top of the lid to maintain the moisture level.

6. **Plan for additional help**
   If you conduct the outdoor search with the whole class, it is helpful to have additional helpers. Consider asking family members or older students to work with small groups of students as they search for snails. Extra eyes will help students search for snails safely.

7. **Plan a snail wake-up call**
   To make the snails active, place them in a 1/2 L container of water or spray them with water from the spray mister just before activity time. (This also cleans them off.) Return them to the terrarium and put the lid on tightly.

8. **Make cardboard fences**
   Fold the cardboard pieces in half the long way so they stand up. They will be used as barriers to block the snails' movements.

9. **Plan for safety**
   Review outdoor safety rules and expectations before going outdoors. Hang up the *Outdoor Safety* poster and review it with students. If there are any areas or plants that are off limits, make sure to remind students and helpers about them.

10. **Check the site**
    Tour the outdoor site on the morning of the activity. Do a quick search for potentially distracting or unsightly items.

11. **Plan to read *Science Resources*: "Water and Land Snails"**
    Plan to read "Water and Land Snails" during a reading period.

12. **Plan assessment for Part 3**
    Plan to listen and observe students as they work at the center. This is a good time to reflect on student engagement with crosscutting concepts.

    What to Look For

    - Students observe where land snails live, how they move, and how land snails change their surroundings. (Constructing explanations; ESS2.E: Biogeology; cause and effect.)

Full Option Science System

Part 3: Land Snails

GUIDING *the Investigation*
Part 3: *Land Snails*

1. **Introduce land snails**
   Call students to the rug. Tell them that you have brought a new animal to class for them to observe. Give them some clues to see if they can figure out its identity.

   This animal has no arms or legs. It moves slowly. It carries its house on its back.

   ➤ *What kind of animal is it?*

   Students should say a snail, but they might say a water snail. Tell students that this snail lives on the land and is called a **land snail**.

   Show students the land snails you collected and the photograph of the land snail on page 30 of the *FOSS Science Resources* book.

2. **Discuss a snail search**
   Tell students that today they will go outside to look for land snails. Ask students for ideas about where snails might live. Encourage students to explain why they think snails live where they suggest. Listen to their ideas and add suggestions of your own. Make a class list of student predictions of where to find snails outdoors.

   Explain that students will work with a partner to find snails. Each pair of students will have a plastic cup and lid to hold the snail when they find one. Every pair should find one snail. Before collecting the snail in the cup, they should observe carefully what the snail is doing. They should remember where they found the snail and what it was doing when they collected it.

3. **Discuss outdoor safety**
   Refer to the *Outdoor Safety* poster and review safety outdoors. Tell students about any plants or areas that should be avoided.

4. **Go outdoors**
   Distribute a cup and lid to each pair of students and head outside to the selected outdoor area. Gather in the sharing circle and describe the boundaries of the search area. Review the expectations for looking for land snails.

   - *Treat all animals and plants with care.*
   - *If you are holding a snail, hold it gently.*
   - *Stay with your partner and within the boundaries of the area.*
   - *When you locate a snail, observe what it is doing. Then collect it and put it in your cup with the lid. Remember where you found the snail.*

FOCUS QUESTION
What do land snails do?

▶ **NOTE**
Caution students to use extra care when handling small snails. Their shells might be somewhat fragile.

Materials for Step 4
- *Plastic cups with lids*

Animals Two by Two Module—FOSS Next Generation

145

INVESTIGATION 2 – Water and Land Snails

> **NOTE**
> Make sure to keep the snails in enclosed cups out of direct sunlight. Closed containers can overheat quickly on a warm day.

SCIENCE AND ENGINEERING PRACTICES
Analyzing and interpreting data

Materials for Step 7
- Plastic cups
- Paper towels
- Land-snail terrarium
- Spray mister

SCIENCE AND ENGINEERING PRACTICES
Planning and carrying out investigations

CROSSCUTTING CONCEPTS
Cause and effect

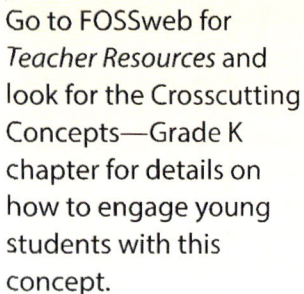

> **NOTE**
> Go to FOSSweb for *Teacher Resources* and look for the Crosscutting Concepts—Grade K chapter for details on how to engage young students with this concept.

5. **Search for snails**
   Have students begin the search. Circulate to groups as they observe and collect the snails.

   When each pair has collected a snail, gather at the sharing circle. Ask a few pairs to share where they found the snails and what the snails were doing.

6. **Return to class**
   Gather the materials and return to class with the collected snails. You can either have students place their snails in the **terrarium** with the other land snails, or go on to the snail observations.

   Review the chart with students' predictions of where they would find snails. Ask students to describe where they actually found snails.

   ➤ Were your predictions correct? What was similar? [Dark, moist, plants, shelter.]

   POSSIBLE BREAKPOINT

7. **Distribute snails**
   At the center, put a snail in a plastic cup for each student. Spray the snails with water to get them moving.

   If you are doing this with the whole class at one time, pairs of students should share a snail.

8. **Observe snail activities**
   Ask students what they could do to learn more about the land snails. They will likely suggest close observations. Tell students they can observe them and handle them carefully.

   After students have observed the snails unguided for several minutes, ask some questions to hone students' observations.

   ➤ How do snails move?
   ➤ What do snails do when you put them in a cup?
   ➤ Do snails move up? Down? **Sideways**?
   ➤ Can snails travel **upside down**? Backward?
   ➤ How does it feel when a snail moves on your hand?
   ➤ What does the snail do when you pick it up by the shell?
   ➤ How can you tell where a snail has been? [It leaves a mucous trail.]

Part 3: Land Snails

9. **Review vocabulary**
   As students offer their observations, add any new or important vocabulary to the class word wall. Let students be the guides—acknowledge the words they use and offer new vocabulary as needed. Record questions students have about snails. Highlight those that can be answered in the investigation.

10. **Make a cardboard fence**
    Give each student at the center (or pair of students if working with the whole class) a piece of folded cardboard to use as a fence or barrier. Let the snails move across the table. Have students guess what will happen when they place their fences in the snails' paths. Then let them observe as the snails confront the barriers.

11. **Observe snail structures**
    Ask students to share with a partner one question they have about snails. Call on a few volunteers to share a question their partner had. Give students time to make observations to answer some of their questions.

    Here are additional questions to motivate close observation of snail structures.

    ➤ *What does the shell look like? Are all the shells the same?*

    ➤ *Do snails have heads? Tails?*

    ➤ *What do you see sticking out from the head?* [Tentacles.]

    ➤ *How many tentacles do you see?* [Four—two long and two short.]

    ➤ *What happens to the tentacles when you touch them gently?*

    ➤ *Do snails have eyes? Ears? A nose? A mouth?*

    ➤ *How do snails breathe?* [On the underside of the snail, close to the shell, there is a hole that opens and closes to let air in.]

    ➤ *Where is the foot of the snail?* [The large surface on which the snail moves.]

    ➤ *How does the foot move?* [Wavelike motions, visible when the snail moves on the surface of a cup.]

    ➤ *How do land snails compare to water snails?*

12. **Record observations (optional)**
    Give each student a copy of the *Land-Snail Outline* sheet to draw and identify the land-snail structures they observe, such as eyes, foot, and tentacles (four tentacles on land snails but only two on water snails). Have them dictate a sentence for the bottom of the sheet or direct them to choose words to add from the class word wall.

---

land snail
sideways
terrarium
upside down

Materials for Step 10
- *Cardboard fences*

SCIENCE AND ENGINEERING PRACTICES

Asking questions

Analyzing and interpreting data

Animals Two by Two Module—FOSS Next Generation

INVESTIGATION 2 – Water and Land Snails

SCIENCE AND ENGINEERING PRACTICES
Constructing explanations

FOCUS CHART

What do land snails do?

Snails move slowly on their foot. They can hang upside down.

They feel with their four tentacles.

They eat leaves.

Materials for Step 15
- *Carrot and lettuce*
- *Spray mister*

▶ **NOTE**
If you collected the snails from some place other than the schoolyard, do NOT release the snails in the schoolyard. If you do not know where the snails were collected or if you purchased them from a biological supplier, let them live out their life in the classroom. Do not release these land snails in the schoolyard.

TEACHING NOTE

See the **Home/School Connection** for Investigation 2 at the end of the Interdisciplinary Extensions section. This is a good time to send it home with students.

---

13. Have a sense-making discussion
Sense-making discussions can be conducted in small groups as part of center time or as a whole class once all students have completed the center activity. The intent is for students to have a sense-making discussion after working with materials and before answering the focus question.

On a large piece of paper, draw a large snail as students watch. Have students take turns sharing snail structures and what that structure does (function) or how the snail uses that structure. If students agree with the description of the structure and function, add it to the drawing.

14. Focus question: What do land snails do?
Write the focus question on the chart as you read it aloud.

➤ *What do land snails do?*

Tell students that you have a strip of paper with the focus question written on it. Describe how to glue the strip into the notebook before they answer the question. When they return to their tables, they should answer the focus question in their notebooks with pictures and words. You might want to model a response on the board or chart paper by making an outline of the snail and providing a sentence frame such as, A snail _____ .

15. Prepare the center for the next group
Have students return their land snails to the terrarium. Collect the cardboard barriers. Wipe out the cups if necessary. Have students wash their hands.

16. Discuss needs of land snails
Review the needs of animals. Ask students what they should do to care for the land snails in the terrarium. Facilitate the discussion to the ideas that the snails need food (carrots and lettuce) and some moisture. Students can spray the snails daily with the spray mister.

Encourage students to observe changes in the terrarium. Ask them how the snails change their own environment.

Work with students to determine what they should do with the land snails they collected from the schoolyard. Suggest to students that they should return them to where they found them. Another alternative is to give them to another class in the school to study.

17. Compare land snails and water snails
Have students continue to observe and compare the water and land snails. Create a class T-table to show how they are different. Make a list of similarities.

Part 3: Land Snails

READING *in Science Resources*

18. Read "Water and Land Snails"

This article extends students' learning of snails on land and in water. Students read about snails that live in water, both fresh water and salt water, and about snails that live on land. The reading provides an opportunity to compare a variety of snails to those in the classroom or outdoors.

Begin by having students share with a partner one thing they have learned about water snails, one thing about land snails, and one question they have about either one. Call on a few volunteers to share their questions and write them on the board or chart paper.

Show students the title of the article and have them read it with you. Ask them what they think they might learn from this reading. Refer to the content grid started in Investigation 1 and review the categories of information the class gathered on fish and birds. Make a new row for "snails" and tell students that this time they should listen and look for information they can use to answer the same questions for snails.

Read aloud the article in the big book, using the strategy that will be most effective for your class. Pause to discuss the main topics and key points in the article, to compare the different types of water and land snails in the photographs, and to respond to the questions in the text.

19. Discuss the reading

Discuss the article, using these questions as a guide. Have students help you locate information in the text and photographs to fill in the information on the content grid.

➤ Where do snails live?

➤ Where is the snail's foot? Shell?

➤ Where are the tentacles? What do they do?

➤ How are the tentacles of land and water snails different?

➤ Which snail looks like the ones we have in our aquarium?

➤ Which ones look like land snails you found outdoors?

Encourage students to ask questions about the snails that might be answered with further investigation in the classroom or in the schoolyard. Add these questions to the list of questions started before the reading. As an extension, go through the list of questions (or pick out a few examples) and talk about how

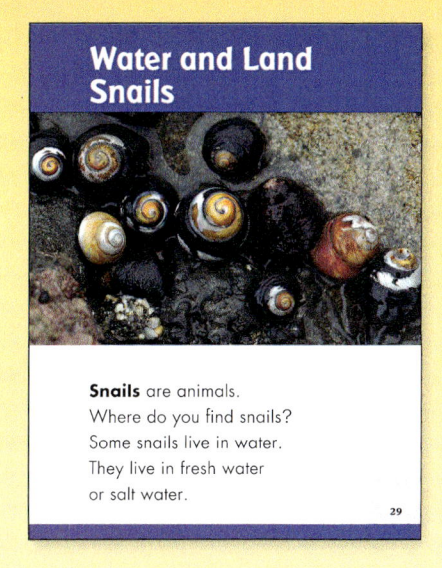

Snails are animals. Where do you find snails? Some snails live in water. They live in fresh water or salt water.

ELA CONNECTION

These suggested strategies address the Common Core State Standards for ELA.

RI 1: Ask and answer questions about key details.

RI 2: Identify main topic and retell key details.

RI 3: Describe the connection between two ideas.

RI 10: Actively engage in group reading activities with purpose and understanding.

W 7: Participate in shared research and writing projects.

W 8: Gather information to answer a question.

SL 3: Ask and answer questions to seek information.

SL 4: Describe with details.

INVESTIGATION 2 – *Water and Land Snails*

SCIENCE AND ENGINEERING PRACTICES

Asking questions

Obtaining, evaluating, and communicating information

CROSSCUTTING CONCEPTS

Patterns

one would go about answering different types of questions. Is this a question we answer by observing snails in the classroom or outdoors? Is this a question we can ask somebody about? Can we find the answer in a book? On the Internet?

20. Share information about photos

Tell students you have more information to share about these animals that might answer their questions. Flip through the pages a second time sharing some of the interesting features of the animals.

Page 29. Black turban snails. These snails are found in the rocky intertidal zone of the ocean. They wedge themselves into cracks of tide pools to protect themselves from predators such as sea otters and sea stars. They eat algae.

Page 30. Brown garden snail. These snails were introduced from Europe but now are found across the United States. They can lay about 80 eggs at a time, up to six times a year.

Page 31. Banded garden snail (inset). Also called a banded wood snail or brown lipped snail; the colorful stripes on the shell are very distinctive. Land snails get water from the plants they eat, rain, and dew. If there is no water around, a land snail can seal up its shell and survive for a long time. Size: 2.5 cm.

Page 32. Roman snail (top). The roman snail is also called the edible snail (escargot). It was brought to the US from Europe to raise and eat. Now it is quite often found in gardens.

Page 34. Ramshorn snail (top). This common aquarium snail has a shell that spirals like the horn of a ram (male sheep). It has two tentacles, while garden snails have four. Can you see the tentacles?

Page 36. Textile cone snail (top left), Giant African land snail (bottom left), flamingo tongue snail (bottom right). The textile cone snail has a beautiful shell, but their hollow teeth will inject poison into their prey. They live in the Indo-Pacific ocean in sandy areas beneath coral and rocks in shallow waters. Size: 9 cm.

The Giant African land snail lives on land and eats so many different plants that it is a serious pest in areas including Florida and Hawaii. It is as big as a person's hand.

The flamingo tongue snail has a colorful pattern covering its body. The pattern is actually on the mantle or soft tissue that wraps around the shell. When the shell is exposed, it is a plain cream color. These snails eat coral. Size: 5 cm.

Part 3: Land Snails

21. Focus on photographs
Have students think about and discuss how they are able to gather information from the photographs in the article. Ask,

➤ *How do the photographs help you understand what the author is trying to say?*

Point out that some of the photographs are not life-sized. Ask,

➤ *Why did the author use big (magnified) images?*

As an extension, students can compare information about snails presented in this article to how snails are described and illustrated in other texts. See FOSSweb for a list of recommended books.

WRAP-UP

22. Share notebook entries
Conclude Part 3 by having students talk with a partner about the guiding question for the investigation. They should use their notebooks as a reference. After sharing with a partner, ask for volunteers to talk about their ideas.

Read the question together.

➤ *What do animals such as snails need to live and grow?*

23. Engage in science talk
This is a good opportunity to engage in science talk. Use one of the questions from the list students generated or ask an interesting question students can discuss based on their observations and/or predictions, such as,

- What types of surfaces do snails prefer?
- How much light do snails prefer?
- What types of food do they prefer?

Refer to the Science Extensions in the Interdisciplinary Extensions at the end of this investigation for more ideas.

Students' ideas can be expressed as claims (what they think) based on evidence (why they think so). Introduce the prompts to students and model how they might be used. You can write the prompts on chart paper as reference for students.

ELA CONNECTION

These suggested strategies address the Common Core State Standards for ELA.

RI 7: Describe the relationship between illustrations and the text.

RI 9: Identify similarities in and differences between two texts on the same topic.

DISCIPLINARY CORE IDEAS

LS1.A: Structure and function
LS1.C: Organization for matter and energy flow in organisms
ESS2.E: Biogeology
ESS3.A: Natural resources

INVESTIGATION 2 – Water and Land Snails

ELA CONNECTION

This suggested strategy addresses the Common Core State Standards for ELA.

SL 1: Participate in collaborative conversations.

SCIENCE AND ENGINEERING PRACTICES

Engaging in argument from evidence

- I think _____ because _____ .
- I agree with _____ because _____ .
- I disagree with _____ because _____ .
- I want to add that _____ .

Remind students to listen to and build on the ideas of others, and to explain their own answers (provide evidence based on their observations and what they've learned about snails from the text).

Interdisciplinary Extensions

INTERDISCIPLINARY EXTENSIONS

Language Extension

- **Keep a classroom snail journal**
  Make one snail journal for the whole class and keep it at the snail center. One student at a time will be able to add observations to the journal. For added interest, cut the front and back cover in the shape of a snail and staple half sheets of writing paper inside.

Math Extensions

- **Use shells for addition and subtraction**
  Give each student a set of ten shells to use in simple addition and subtraction problems. For example, have students make a group of two shells and a group of three shells. Have them put the groups together and count to see how many they have altogether. Or begin with all ten shells. Have students take away four and tell how many are left.

- **Make equal sets**
  Put five to ten shells in plastic bags, one bag for each student in the group. Put a different number of shells in each bag. Ask students if they think the shells have been divided fairly. Have them share shells until everyone has the same number.

- **Divide the shells**
  Give 12 shells to each student in a small group. Ask students to divide their set of shells into two equal groups, three equal groups, four equal groups, or six. Discuss their method for making equal groups and how many shells ended up in each group.

- **Use shells to make patterns**
  Set up a few patterns for students to complete using the shells. Have them begin a pattern and have another student complete it.

- **Measure how far a snail moves in 1 minute**
  Measure how far a snail moves in 1 minute (or 2 minutes) by taping the end of a spool of thread to the shell with duct tape and letting it unwind as the snail moves. Put the spool of thread on a toothpick and hold it in place as the snail moves. Cut the thread when time is up.

> **TEACHING NOTE**
>
> *Refer to the teacher resources on FOSSweb for a list of appropriate trade books that relate to this module.*

Animals Two by Two Module—FOSS Next Generation

153

INVESTIGATION 2 – *Water and Land Snails*

Art Extension

- **Make paper-bowl snails**

  To make the snail pictured in the sidebar you will need two paper bowls, a snail template, and one pipe cleaner for each student. Prepare the materials.

  - Draw the cut lines on the bottom of two paper bowls per student. The cut lines will cut a flat edge on the two bowls to form the bottom of the shell next to the foot.
  - Use teacher master 17, *Land-Snail Body*, to make snail templates out of tagboard for students to trace around.
  - Cut pipe cleaners into one 10 centimeter (cm) and one 5 cm section for each student.

  Demonstrate snail construction.

  - Cut the bowls on the line drawn to make a straight edge.
  - Trace and cut out of construction paper a snail body.
  - Sandwich the bowls around the snail body, and staple into place.
  - Use the 10 cm pipe cleaner to punch through the snail body and make long tentacles for the eyes at the top of the head; use the 5 cm pipe cleaner for the "feeler" tentacles that are down lower. Pinch the pipe cleaners together at the base of the head to hold them in place.
  - Use crayons, markers, or paints to draw the spiral on the shell.

Science Extensions

- **Test for surface preferences**

  Set up an investigation to test for snail preferences. For example, students might put an active snail between a piece of foil and a piece of waxed paper and see which way it moves. Test several times to see if it chooses the same path each time and truly has a preference. You might also try

  - sandpaper versus construction paper
  - moist versus dry surfaces
  - warm versus cool surfaces
  - a light versus a dark area
  - food preferences

  The snails might *not* show a preference in all cases. "No preference" is an important concept to begin developing through these activities.

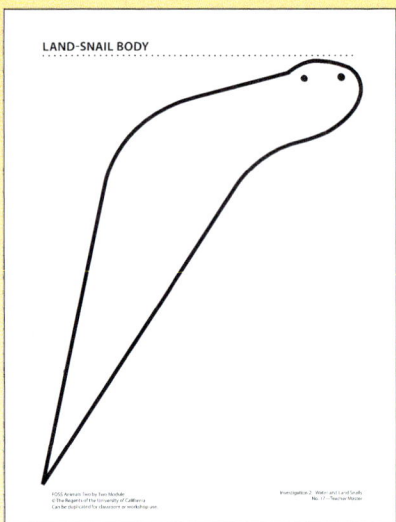

No. 17—Teacher Master

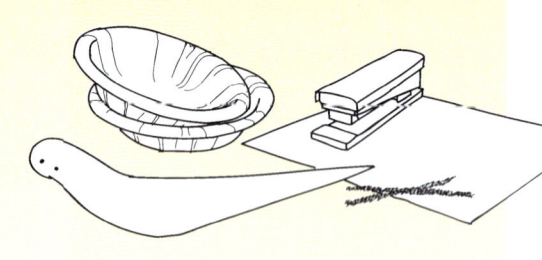

TEACHING NOTE

Review the online activities for students on FOSSweb for module-specific science extensions.

Interdisciplinary Extensions

- **Are snails attracted to direct light?**
  Have students plan and conduct a simple investigation to find out if snails are attracted to light. Students can shine a flashlight to see if the snails move toward it or away from it.

- **Conduct land snail races**
  Hold land snail races. A land snail race track comes in the kit as two separate laminated sheets, 28 cm × 43 cm. Tape the two racetrack sheets together with transparent tape.

  Bring out the race track and place it in the center of the table. Explain that the race will be from the center circle to the outer circle. The snails can go in any direction.

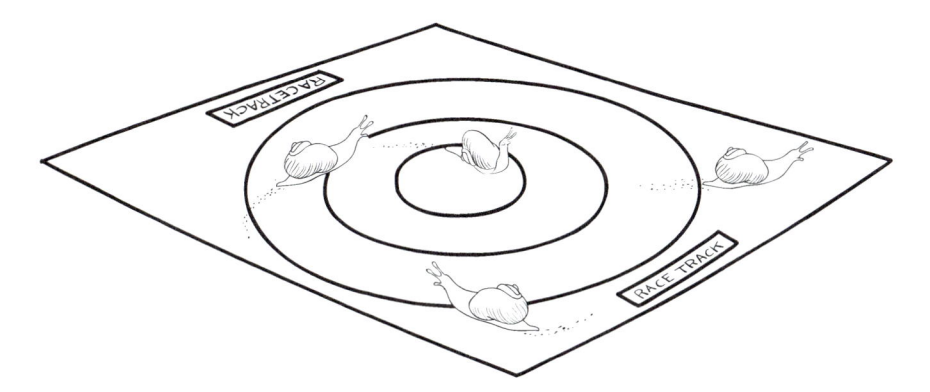

  Have all the students in the group place their snails in the center circle. As soon as the snails are in place, begin the race. Each snail that crosses the outermost circle is a winner.

  Ask questions to guide discussion at the end of one race.

  ➤ Are big snails faster than small snails?
  ➤ Why do you think some snails are faster than others?
  ➤ Were you able to get your snail to go where you wanted it?
  ➤ What could you do to make your snail go faster?

  Hold several races. Let students try encouraging the snails with food or other means to get them to move faster toward the outer circle.

- **Compare slugs to snails**
  Bring in some slugs and compare them to the snails. If students ask, slugs are not snails that have lost their shells; they are a separate species of mollusks. They live their entire lives without a shell.

> **NOTE**
> If all the snails are estivating, rinse them with water to get them moving.

INVESTIGATION 2 – *Water and Land Snails*

> **NOTE**
> Even though the food coloring is non-toxic, rinse the snails off before returning them to the terrarium.

TEACHING NOTE
Encourage students to use the Science and Engineering Careers Database on FOSSweb.

No. 18—Teacher Master

- **Observe snail trails outdoors**
  If snails are abundant in your schoolyard, take students outside in the early morning to look for the cellophane-like mucous trails left by snails that have traveled across sidewalks at night.

- **Observe snail trails indoors**
  Have students dip the foot of a snail in water colored with food coloring. Have them set the snail on a piece of white paper and observe the trail the snail leaves.

Environmental Literacy Extension

- **Take a trip to the seashore with naturalists**
  Take students on a virtual trip to the seashore with naturalists who study the mangrove and underwater grass bed ecosystems. Students will learn about collection and observation techniques and see a diversity of fish and mollusks close-up. They will also hear from scientists about how humans can benefit from understanding the organisms in an ecosystem.

  To start the field trip, go to the streaming videos section on FOSSweb *Seashore Surprises,* Chapters 4 and 5, "Seashore Habitats."

Home/School Connection

Now that students have had several "up close" experiences watching animal behaviors, they are primed to play a game of animal "charades." At home, students lead their families in a game of "Who Am I?" acting out the behaviors of different animals.

Print or make copies of teacher master 18, *Home/School Connection* for Investigation 2, and give these to students to take home after Part 3.

As an additional home/school connection, ask students to talk to their families about what they have learned about snails. Students can also share any family stories or artifacts related to snails and/or shells with the class.

INVESTIGATION 3 – Big and Little Worms

Part 1
The Structure of
Redworms 168

Part 2
Redworm Behavior.............. 174

Part 3
Comparing Redworms to
Night Crawlers 182

Guiding question for phenomenon:
What do animals such as worms need to live and grow?

Science and Engineering Practices
- Asking questions
- Developing and using models
- Planning and carrying out investigations
- Analyzing and interpreting data
- Constructing explanations
- Engaging in argument from evidence
- Obtaining, evaluating, and communicating information

Disciplinary Core Ideas
LS1: How do organisms live, grow, respond to their environment, and reproduce?
LS1.A: Structure and function
LS1.C: Organization for matter and energy flow in organisms
ESS2: How and why is Earth constantly changing?
ESS2.E: Biogeology
ESS3: Earth and human activity
ESS3.A: Natural resources

Crosscutting Concepts
- Patterns
- Cause and effect
- Systems and system models
- Structure and function

PURPOSE

Students have firsthand experiences with two related animals—redworms and night crawlers. Through observation and discussion, students gather information about earthworm structures and behaviors and how those characteristics relate to the needs of the animals. Students focus on the phenomenon of earthworms as recyclers of plant material.

Content

- Worms are animals and have basic needs—water, food, air, and space with shelter.
- Worms have identifiable structures. Different kinds of worms have similar structures and behaviors; they also have differences (size, color).
- Worm behavior is influenced by conditions in the environment. Worms change plant material into soil.

Practices

- Observe and compare the structures and behaviors of redworms and night crawlers.
- Compare and communicate how redworms and night crawlers are the same and different.

FOSS Full Option Science System

INVESTIGATION 3 – Big and Little Worms

	Investigation Summary	Time	Focus Question for Phenomenon, Practices
PART 1	**The Structure of Redworms** Students dig through a terrarium to discover that there are redworms living in the soil. They look for some of the structures they have seen on other animals they have studied so far. They rinse the worms in water to remove the soil and to get a better view.	**Introduction** 5 minutes **Center** 15–20 minutes **Notebook** 15 minutes	**What are the parts of a redworm?** **Practices** Planning and carrying out investigations Analyzing and interpreting data
PART 2	**Redworm Behavior** Students focus on the movement and behavior of redworms. They notice how the worm's body contracts and stretches to move forward. They observe the worm to see if it can move in other directions. They try blocking the worm's path to see what it does. Students consider what redworms need to live and set up a worm-jar habitat. Students observe how redworms change leftover food and other plant materials into soil.	**Introduction** 5 minutes **Center** 15–20 minutes **Notebook** 15 minutes	**What do redworms need to live?** **Practices** Asking questions Developing and using models Planning and carrying out investigations Analyzing and interpreting data Constructing explanations
PART 3	**Comparing Redworms to Night Crawlers** Students discover a new kind of worm in their terrarium—night crawlers. The new worms are much longer and fatter than the redworms. Students observe the two kinds of worms and compare the structures and behaviors of the two animals.	**Introduction** 5 minutes **Center** 15–20 minutes **Notebook** 10–15 minutes **Reading** 15 minutes	**How are redworms and night crawlers different?** **How are they the same?** **Practices** Asking questions Planning and carrying out investigations Analyzing and interpreting data Constructing explanations Engaging in argument from evidence Obtaining, evaluating, and communicating information

At a Glance

Content Related to DCIs	Writing/Reading	Assessment
• Worms have identifiable structures. • Worms are animals and have basic needs—water, food, air, and space with shelter.	**Science Notebook Entry** *Worm Outline*	**Embedded Assessment** Teacher observation
• Worms are animals and have basic needs—water, food, air, and space with shelter. • Worm behavior is influenced by conditions in the environment. • Worms change plant material into soil.	**Science Notebook Entry** *Worm-Jar Outline*	**Embedded Assessment** Teacher observation
• Different kinds of worms have similar structures and behaviors; they also have differences (size, color).	**Science Notebook Entry** Draw or write words to answer the focus questions. **Science Resources Book** "Worms in Soil"	**Embedded Assessment** Teacher observation **NGSS Performance Expectations addressed in this investigation** K-LS1-1 K-ESS2-2 K-ESS3-1

Animals Two by Two Module—FOSS Next Generation

INVESTIGATION 3 – Big and Little Worms

BACKGROUND *for the Teacher*

Worms are a varied lot. You might have heard of roundworms, flatworms, tapeworms, earthworms, and who knows what other kinds of worms. None of them conjures up a particularly warm or pleasant feeling in most people. Worms have low reputations in human circles, often associated with some not-so-pleasant circumstances. But this activity might turn all that around as you dig into the phenomenon of earthworms.

What Are the Parts of a Redworm?

Earthworms are members of the phylum *Annelida*, or ringed animals. They are fairly simple life-forms, put together from a number of disklike **segments** like a long flexible roll of coins. The worm **body** has no internal skeleton like a fish, no hard protective exoskeleton like an insect, and no shell into which they can withdraw. Worms are flexible, elongated bundles of muscle, efficiently designed for life in compost or underground.

The characteristic wriggling of earthworms is accomplished by the contraction of two kinds of muscles. When the short muscles that circle each segment (like lots of rings on a finger) contract, the worm gets thinner and longer. When the long muscles that connect all the segments contract, the head and tail are pulled toward each other, and the worm becomes short and fat. Depending on which end of the worm is anchored, the worm can move along the surface of the ground or through its burrow effectively in either direction, head first or tail first.

These are the physical structures of the **redworm** that students will observe—they will see the pointed worm head end, which is red and slightly larger than the pointed tail end, often lighter in color. The worm mouth is at the head end. Students will see the segments as the worm expands and contracts as it moves.

Earthworm organs are quite different from ours, making it possible for them to live their very different lifestyle efficiently. Earthworms have five pairs of simple hearts that pump blood throughout the body. They have no lungs. Instead, the blood flowing close to the worm's surface absorbs oxygen and releases carbon dioxide directly through the **moist** skin (called the cuticle). For this reason, earthworms can live for some time in water if the oxygen supply is adequate. They don't drown per se, but they might suffocate if the oxygen content is low. This is why worms leave the **soil** and crawl out on the sidewalk during a heavy rain—they are seeking oxygen. Earthworms are not adapted to feed in water, however, so they would starve to death in due course.

Background for the Teacher

Instead of a nose, ears, and eyes, earthworms have a nervous system throughout their bodies that controls actions in response to environmental stimuli, such as vibrations, heat, cold, moisture, light, and the presence of other worms. However, they have no brain, so worms do not ponder their lowly lot in life, nor do they plan a strategy for obtaining their next meal or crossing the sidewalk safely.

Like all animals, earthworms have effective strategies for begetting their own kind. With earthworms, it is not a matter of boy meets girl, but rather a simpler matter of worm meets worm. All worms produce both eggs and sperm, but they cannot fertilize their own eggs—mating is still a necessary part of reproduction. Mature earthworms have an enlarged or **swollen** band some distance from the head (about one-third of the distance between the head and the tail). This enlarged **clitellum** plays an important role in reproduction. The clitellum is more subtle and harder to see in the redworm, but students will observe it in the larger **night crawler**.

In mating, two worms approach each other nose to nose. With their bodies touching, they slide past each other until their heads are a bit past the clitellum. Both worms pass sperm through an opening located between the head and the clitellum, into a temporary holding receptacle in the other worm. The two worms separate. The clitellum secretes a liquid that solidifies into a flexible tube. As the tube lengthens, the worm backs out of it. Soon the tube covers the front part of the worm. The worm lays a few eggs inside the tube, deposits some of the stored sperm, and withdraws from the tube, leaving the eggs and sperm inside the tube. The ends of the tube pinch off to form a cocoon, and the whole thing shrinks to a tidy package about the size of a fat grain of rice. The cocoon is left alone, sitting on or just under the surface of the soil. The worm continues to produce cocoons until the sperm is used up. Cocoons are durable, can overwinter in cold climates, and can wait out hot dry spells in arid environments. After 3 weeks (ideal conditions) or longer, the cocoon opens, and out sallies the next generation.

Earthworms feed on decomposing organic material, mostly vegetation, from the surface of the soil and within the soil itself. In the process of burrowing and feeding, they process tons of soil in a typical pasture or garden, improving the quality of soil for plants and other animals. There are some 1800 species of earthworms worldwide. Some are tiny, no more than 2 centimeters (cm) at maturity. At the other end of the scale are

"Instead of a nose, ears, and eyes, earthworms have a nervous system throughout their bodies that controls actions in response to environmental stimuli..."

Animals Two by Two Module—FOSS Next Generation

INVESTIGATION 3 – Big and Little Worms

the Australian giants that average about 2.5 meters (m) in length, and the record holder, a South African gargantuan measuring 7 m in length. Not to worry—the largest earthworms in North America are the common night crawlers, which can reach a length of little more than 30 cm. And the earthworms that you will use in this investigations are half that length or shorter.

What Do Redworms Need to Live?

The redworm or red wriggler (*Eisenia fetida*) is relatively small (3–8 cm long) and very thin. Although they are small, they can eat nearly half of their weight in food every day. They eat decaying leaves and other decaying plant parts that have been broken down. This is the common species of redworm used in vermicomposting. In worm bins, redworms eat grass clippings and kitchen scraps, including vegetables, fruits, egg shells, coffee grounds, paper, and cardboard.

Redworms are earthworms known as epigeic worms—that is to say, they live on the surface of the soil or in the **top** 25 cm or so of the topsoil under the litter layer. They are specialized for life in rapidly changing environments, such as compost piles. They prefer darkness and can tolerate the warmer conditions and high worm population density found in leaf litter and compost. While some earthworms live in permanent, deep, vertical, cool burrows in solitude, redworms live in large populations and burrow randomly through the litter layer on top of soil.

Redworms breed and consume wastes much faster than other earthworms known as soil worms, or anecic worms. The night crawlers (*Lumbricus terrestris*) used in Part 3 are anecic worms—they build permanent, vertical burrows that penetrate the soil deeply. Night crawlers are detritivores and come to the surface to feed on partially decomposed litter, manure, and other organic matter. These earthworms can often be found deep in their burrows where the temperature is relatively stable and much cooler than in the leaf litter on the soil surface.

Redworms have a rudimentary digestive system. They have no teeth and few digestive fluids, so they depend on fungi and bacteria to predigest their food. Redworms can survive on any organic matter, including cellulose and starches, as long as it is broken down by the action of air and water and microorganisms to break down the fibers. Redworms also have some of these organisms in their intestines. It is the microbes that actually digest the food and turn it into nutrient-rich worm castings that are the desired results of vermicomposting.

Background for the Teacher

Use dry leaves that are already decomposing to feed the worms. There are a few types of leaves that you should not use as food for the worms—bay, eucalyptus, and magnolia leaves, and needles from pine, fir, and cedar trees. These types of leaves will kill your worms.

A good temperature for redworms is 15° to 26°C—good room temperatures. (Night crawlers need much cooler temperatures and will do fine in the refrigerator.) Redworms will slow down their activities in colder or hotter temperatures.

Redworms require moisture to breathe because they take in oxygen through their skin. If they dry out, they will die. But too much moisture is not good for them either. A spray mister is a good way to provide moisture in their worm jar. Shredded newspaper is a good way to moderate the moisture in worm jars.

The redworms can also be kept in a small terrarium. The soil should be moist, but without any pools of standing water. Shredded newspaper is useful as bedding to create spaces and texture in the soil for the worms. Use the mister to spray the surface of the terrarium once a day, and lay a sheet of clear plastic wrap loosely over the lid of the terrarium to maintain a moist environment. You do not need to dechlorinate the water that is used in the terrarium. Add to the soil a small amount of oatmeal for food and lay a few decaying leaves on the surface of the soil.

▶ **NOTE**
As a policy, you should never release organisms into natural environments. Redworms (*Eisneia sp.*) are not found in natural environments, but only in human environments (compost bins, horse manure piles). For that reason, you can add redworms to your worm farm or compost bin at the end of this investigation. Night crawlers should not be released into natural environments.

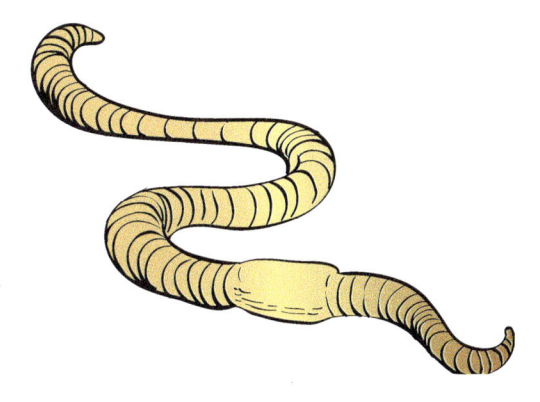

How Are Redworms and Night Crawlers Different? How Are They the Same?

The night crawlers are also earthworms with segmented bodies, a head end, and a tail end; they move in the same way as redworms. Because night crawlers are larger, the clitellum is more obvious. Students will be able to feel the **bristles** on the ventral (belly) side of this earthworm. Night crawlers prefer cooler temperatures and deep burrows so will not survive well in the terrarium with redworms over long periods of time. You can maintain night crawlers in a container with some moist, loose soil and food. Unlike redworms, night crawlers are hard to keep alive in the classroom and will soon die and become part of the soil. If you want to keep them alive for a month or so, keep them at a cool temperature by putting them in the warmest part of the refrigerator.

Animals Two by Two Module—FOSS Next Generation

INVESTIGATION 3 – Big and Little Worms

TEACHING CHILDREN about Worms
Developing Disciplinary Core Ideas (DCI)

> **NGSS Foundation Box for DCI**
>
> **LS1.A: Structure and function**
> - All organisms have external parts. Different animals use their body parts in different ways to see, hear, grasp objects, protect themselves, move from place to place, and seek, find, and take in food, water, and air. Plants also have different parts (roots, stems, leaves, flowers, fruits) that help them survive and grow. (foundational)
>
> **LS1.C: Organization for matter and energy flow in organisms**
> - All animals need food in order to grow. They obtain their food from plants or from other animals. Plants need water and light to live and grow. (K-LS1-1)
>
> **ESS2.E: Biogeology**
> - Plants and animals can change their environment. (K-ESS2-2)
>
> **ESS3.A: Natural resources**
> - Living things need water, air, and resources from the land, and they live in places that have the things they need. Humans use natural resources for everything they do. (K-ESS3-1)

It is important that students have plenty of time to investigate earthworms on their own terms. Some will be reluctant at first to touch the worms because of a misconception that they are dirty. (Well, they do have dirt on them, so we should honor the difficulty and confusion students will have if they have been admonished not to get dirty.) Some students will be more receptive to the idea of handling washed worms. Washing is good for other reasons as well—it keeps the worms wet, and it makes it much easier to see what the worm looks like in detail.

Earthworms are animals that live their lives almost entirely under the surface of the ground litter or in burrows under the ground—hidden organisms. The experiences students have in this investigation will start them along the pathway of understanding the rich proliferation of life that makes the whole Earth alive. Every conceivable corner and back alley of the environment is home to some life-form. The experiences with earthworms will expand students' understanding of the ways animals can live and that there are different kinds of worms within the category of earthworms.

While investigating earthworms, students will observe and describe new structures and behaviors. They might report lines around the worms. This is a good time to introduce the word *segment* in a logical context. "Mandy sees lines on her worm. It looks like the worm is made from lots of little pieces stuck together. Can you see the lines and segments on your worm? Do all the worms have segments? Are all the segments the same size?"

Students will also see the worms digging into the soil and moving through "tunnels" under the surface of the soil in the observation terrarium and the worm jars. The underground tunnels are called burrows, and the process of digging is burrowing. Again, seize the opportunity to develop the language in a naturalistic context.

Earthworms are fairly sturdy creatures, but students should be cautioned to handle them with the care accorded any living thing. Encourage students not to tug too hard on the worms or drop them. If an earthworm becomes severed, it can sometimes regenerate, so don't throw it away. Put it in a separate container of moist soil and see what happens.

Teaching Children about Worms

The activities and readings students experience in this investigation contribute to the disciplinary core ideas **LS1.A, Structure and function: All organisms have external parts; LS1.C, Organization for matter and energy flow in organisms: All animals need food in order to grow; ESS2.E, Biogeology: Plants and animals can change their environment;** and **ESS3.A, Natural resources: Living things need water, air, and resources from the land, and they live in places that have the things they need.**

Engaging in Science and Engineering Practices (SEP)

In this investigation, students engage in these practices.

- **Asking questions** about redworms habitats.
- **Developing and using models** by drawing the redworm habitat over time to represent patterns in the natural world.
- **Planning and carrying out investigations** with redworms to observe their structures and study their environmental needs.
- **Analyzing and interpreting data** by describing observations of the redworms over time, recording information, using and sharing notebook entries, including writing and labeled pictures. Students use their firsthand observations and those of others in the classroom to describe the patterns they observe in the redworms habitats.
- **Constructing explanations** by making firsthand observations of redworms and night crawlers and using this as evidence to answer questions about the needs of animals, including food.
- **Engaging in argument from evidence** to support a claim about worms and their importance to soil.
- **Obtaining, evaluating, and communicating information** about structures of two kinds of worms, their needs, and where they live.

> **NGSS Foundation Box for SEP**
>
> - **Ask questions** based on observations to find more information about the natural and/or designed world(s).
> - **Use a model** to represent relationships in the natural world.
> - **With guidance, plan and conduct an investigation** in collaboration with peers (for Grade K).
> - **Make observations** (firsthand or from media) and/or measurements to collect data that can be used to make comparisons.
> - **Make predictions** based on prior experiences.
> - **Record information** (observations, thoughts, and ideas).
> - **Use and share pictures, drawings, and/or writings** of observations.
> - **Use observations (firsthand or from media)** to describe patterns in the natural world in order to answer scientific questions.
> - **Compare predictions** (based on prior experiences) to what occurred (observable events).
> - **Make observations** (firsthand or from media) to construct an evidence-based account for natural phenomena.
> - **Construct an argument** with evidence to support a claim.
> - **Read grade-appropriate text** and/or use media to obtain scientific and/or technical information to describe patterns in the natural world.
> - **Communicate** information or solutions with others in oral and/or written forms using models and/or drawings that provide detail about scientific ideas.

INVESTIGATION 3 – Big and Little Worms

> **NGSS Foundation Box for CC**
> - **Patterns:** Patterns in the natural and human designed world can be observed, used to describe phenomena, and used as evidence.
> - **Cause and effect:** Events have causes that generate observable patterns. Simple text can be designed to gather evidence to support or refute student ideas about causes.
> - **Systems and system models:** Objects and organisms can be described in terms of their parts. Systems in the natural and designed world have parts that work together.
> - **Structure and function:** The shape and stability of structures of natural and designed objects are related to their function(s).

Exposing Crosscutting Concepts (CC)

In this investigation, the focus is on these crosscutting concepts.

- **Patterns.** Structures of two kinds of earthworms—redworms and night crawlers—are similar but they differ in appearance (size, color) and where they live.
- **Cause and effect.** Earthworms can change their environment over time, turning food and paper into soil.
- **Systems and system models.** Earthworms can be described in terms of their structures.
- **Structure and function.** The observable structures of earthworms (head, tail, segments, clitellum) serve functions in survival.

Connections to the Nature of Science

This investigation provides connections to the nature of science.

- **Scientific investigation use a variety of methods.** Scientific investigations begin with a question. Scientists use different ways to study the world.
- **Scientific knowledge is based on empirical evidence.** Scientists look for patterns and order when making observations about the natural world.

New Word
Say it • See it • Hear it • Write it

Body
Bristle
Clitellum
Earthworm
Moist
Night crawler
Redworm
Segment
Soil
Swollen
Top

Teaching Children about Worms

Conceptual Flow

Students continue to explore how animals of all kinds have various needs to live and grow. To understand an animal's needs, you need to first get to know the animal—its structures and behaviors. The third group of animals students study and care for live primarily on land. Earthworms are the main phenomenon in this experience. The guiding question is what do animals such as worms need to live and grow?

The **conceptual flow** for this investigation starts with an introduction to an earthworm—a small segmented redworm. Students dig for worms in the terrarium and then observe and describe their structures. Students can compare their structures to fish and snails.

In Part 2, students observe the behavior of redworms. They set up a worm jar with moist newspaper strips, food scraps, and other plant material. They consider how the jar and its contents provide for the basic needs of the worms. Over a few weeks, students observe the changes that take place in the jar, caused by the action of the worms eating the food and turning it into soil.

In Part 3, students are introduced to a large earthworm, the night crawler. Students compare the structures and behaviors of the large worm to the smaller redworm.

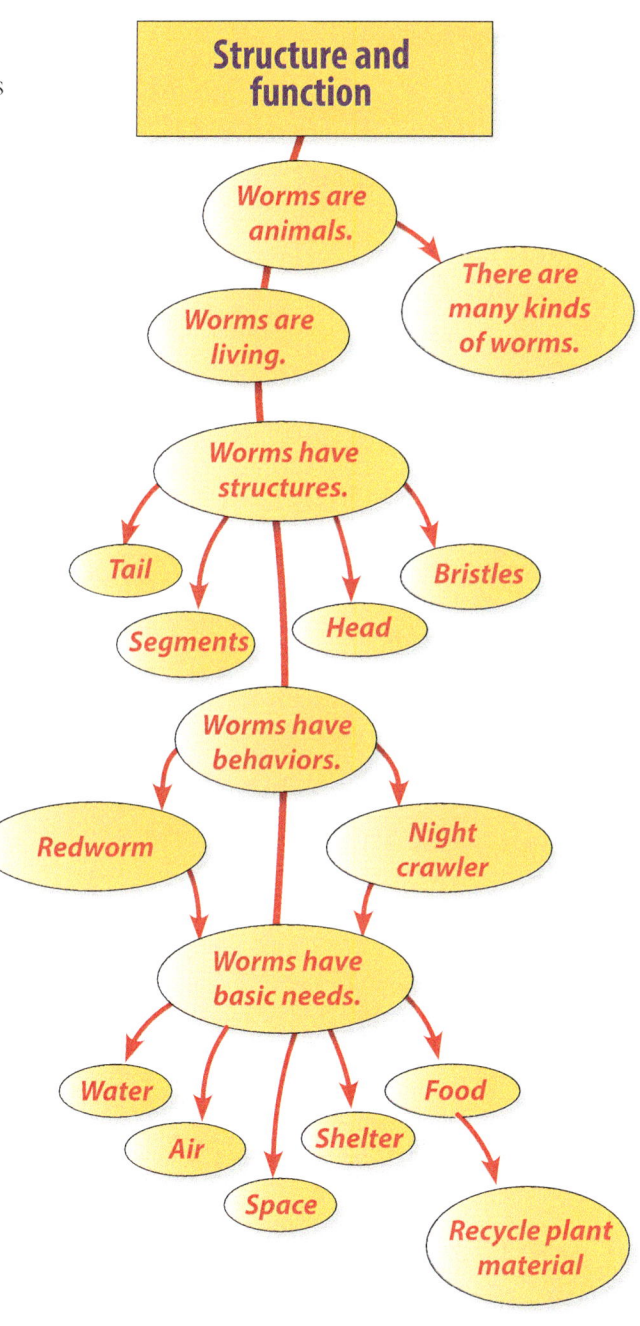

Animals Two by Two Module—FOSS Next Generation

167

INVESTIGATION 3 – Big and Little Worms

MATERIALS for
Part 1: *The Structure of Redworms*

For each student at the center

  1 Plastic cup
  1 *Worm Outline* ★
  1 Hand lens

For the class

  1 Clear basin with lid
  • Plastic wrap
  • Soil, about 2 L
  1 Container and lid, 1/2 L
  1 Plastic cup
150 Redworms (See Step 3 of Getting Ready.) ★
 12 Night crawlers (See Step 3 of Getting Ready.)★
  • Leaf litter ★
  • Oatmeal ★
  1 Pitcher or water container ★
  • Water ★
  • Paper towels ★
❏ 1 Teacher master 19, *Center Instructions—The Structure of Redworms*
❏ 1 Teacher master 20, *Worm Outline*

For assessment

❏ • *Assessment Checklists* 1 and 2

★ Supplied by the teacher.   ❏ Use the duplication master to make copies.

No. 19—Teacher Master

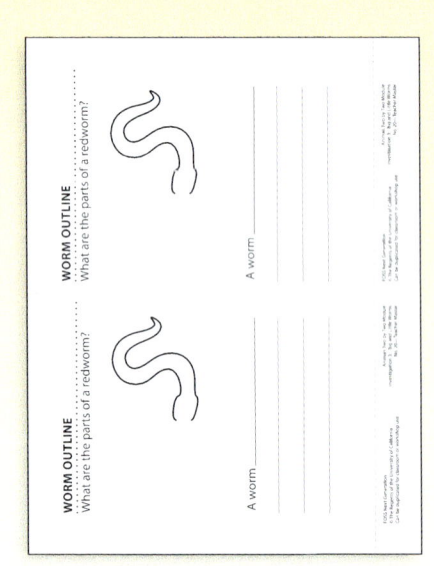

No. 20—Teacher Master

Full Option Science System

Part 1: The Structure of Redworms

GETTING READY for
Part 1: The Structure of Redworms

1. **Schedule the investigation**
   This part requires 15–20 minutes at the center for each group of six to ten students. Plan 5 minutes for an introduction and 15 minutes for students to write and draw in their notebooks.

2. **Preview Part 1**
   Students dig through a terrarium to discover that there are redworms living in the soil. They look for some of the structures they have seen on other animals they have studied so far. They rinse the worms in water to remove the soil and to get a better view. The focus question is **What are the parts of a redworm?**

3. **Obtain worms**
   Bait shops usually carry different kinds of earthworms for fishing. Purchase about 150 redworms and 12 night crawlers. Or you can order worms from Delta Education (see the Materials chapter).

4. **Set up redworm terrarium**
   Use a 1/2 liter (L) container to put three scoops of garden soil or potting soil in a clear basin. Moisten the soil, but don't make it wet. Put the redworms in the terrarium. Add a sprinkling of oatmeal for food and a few decaying leaves on top of the soil.

   There are a few types of leaves that you should not use as food for the worms—bay, eucalyptus, and magnolia leaves, and needles from pine, fir, and cedar trees. These types of leaves will kill your worms.

5. **Prepare night crawlers**
   Keep the night crawlers in the container they came in, or a 1/2 L container about three-quarters full of damp soil. Poke holes in the lid for ventilation. Store the worms in the refrigerator or a cool place for a few days before they are introduced to students.

▶ **NOTE**
To prepare for this investigation, view the teacher preparation video on FOSSweb.

▶ **NOTE**
Redworms are about 6 cm long; night crawlers are 15–25 cm long.

Animals Two by Two Module—FOSS Next Generation

INVESTIGATION 3 – Big and Little Worms

6. **Plan assessment for Part 1**

    Plan to listen and observe students as they work at the center and to review their notebook entries during or after class. Record your observations on *Assessment Checklists* 1 and 2.

    What to Look For

    - *Students understand that animals (redworms) have external parts that help them to meet their needs. (LS1.A: Structure and function.)*

    - *Students describe observations of the structures of redworms, record information by writing and labeling pictures, and share their notebook entries with others. (Analyzing and interpreting data; structure and function.)*

Part 1: The Structure of Redworms

GUIDING *the Investigation*
Part 1: *The Structure of Redworms*

1. **Introduce redworms**
   Call students to the rug. Show them the terrarium. Ask them to guess what kind of animal is living in the **soil**. Tell them that these are very valuable animals because they help keep soil healthy for plants to grow in. If they haven't already guessed, tell them that you have some redworms (a kind of earthworm) living in the soil for them to investigate to find out what they need to live and grow.

2. **Focus question: What are the parts of a redworm?**
   Introduce the focus question on the chart as you read it aloud.

   ▶ *What are the parts of a **redworm**?*

   Remind students that they will again be working with living organisms that must be held gently so they won't be injured.

   Ask students to talk with a partner about what they could do to find out more about how the redworms live. Ask for a few volunteers to share their ideas. If they don't suggest it, explain that they will need to dig through the soil to find the worms.

   Show them how to gently push the soil aside to find the worms. Pull one out and place it in the palm of your hand. After students have had a moment to look at the worm, return it to the terrarium, gently placing it on top of the soil. Ask if students have any questions, then send six to ten students to the learning center.

3. **Review handling redworms**
   When students first arrive at the center, tell them,

   Earthworms have very soft bodies, so be careful not to squeeze them. Their bodies feel **moist**, but they are not slimy.

4. **Dig for worms**
   Give each student a plastic cup. Pass the terrarium around and let two or three students dig at a time to find an earthworm to observe. Remind students to push the soil aside gently so they don't injure any worms. When they find a worm, they can hold it in their palm or put it in the cup.

   If you have a few students who are reluctant to dig for worms, let them watch the other students. They will most likely choose to join in after watching the fun their peers have.

FOCUS QUESTION
What are the parts of a redworm?

New Word — Say it, See it, Hear it, Write it

SCIENCE AND ENGINEERING PRACTICES
Planning and carrying out investigations

New Word — Say it, See it, Hear it, Write it

Materials for Step 4
- *Redworm terrarium*
- *Plastic cups*
- *Container of water*
- *Paper towels*

Animals Two by Two Module—FOSS Next Generation

171

INVESTIGATION 3 – Big and Little Worms

TEACHING NOTE

The dunking usually makes the worms a little more active. Do not let worms remain underwater, however, as they will drown. And be careful not to drop the worms in the process.

CROSSCUTTING CONCEPTS

Systems and system models
Structure and function

body
clitellum
earthworm
moist
redworm
segment
soil
swollen
top

SCIENCE AND ENGINEERING PRACTICES

Analyzing and interpreting data

5. Rinse worms

Pour about 2.5 cm of water into each student's cup to rinse off the worms. Have students remove the worms from the water and place them on the table or in their hands. Let them continue to hold and observe worms without guidance as long as interest lasts. Have students use hand lenses to observe the worms.

6. Guide observations

After several minutes of unguided observations, ask questions to focus student observations.

➤ *Which end is the head and which end is the tail? How do you know?*

➤ *What is the worm's head doing?*

➤ *Look at the color of the worm's **body**. Is it the same in all places?*

➤ *Can you tell the **top** side of the earthworm from the bottom side?*

➤ *Look at the rings (**segments**) that make up the worm's body. Are all the segments the same size? How many do you think there are? Do they all look the same?*

The large **swollen** area is called the **clitellum**, meaning "saddle." It might not be easily seen on the small redworms.

➤ *Does the worm have a mouth?* [Yes, but it is hard to see.]

➤ *Does the worm have eyes? Ears? A nose?* [None that we can see.]

7. Review vocabulary

As students offer their observations, add any new or important vocabulary to the class word wall. Let students be the guides—acknowledge the words they use and offer new vocabulary as needed.

8. Answer the focus question

Wipe off sections of the table and give each student a copy of teacher master 20, *Worm Outline*, to draw the structures they observe, such as segments and head. The focus question is on the sheet. Or, have students draw directly into their notebook without the sheet as a scaffold. Have students dictate a sentence for the bottom of the sheet or direct them to choose words to add from the class word wall.

Ask them to return to their tables and work in their notebooks.

Part 1: The Structure of Redworms

9. **Feed worms**
   Have students put the earthworms back in the terrarium. Sprinkle a little water on top for the redworms to drink. Let each student add a pinch of oatmeal to the terrarium for the worms to eat.

10. **Prepare the center for the next group**
    Empty the cups of water (be careful not to wash a lot of soil down the sink) and stack them so they'll be ready for the next group. Wipe off the table if necessary. Have students wash their hands.

WRAP-UP/WARM-UP

11. **Share notebook entries**
    Conclude Part 1 or start Part 2 by having students share notebook entries. Ask students to open their science notebooks to the most recent entry. Read the focus question together.

    ➤ *What are the parts of a redworm?*

    Ask students to pair up with a partner to
    - share their answers to the focus question;
    - explain their drawings.

    Remind students how to critique their own work. Have them share with a partner one thing they like about their entry and one thing they can make better. Give them a few minutes to revise their entries by labeling the worm parts using the words from the word wall, making a detailed drawing of the redworm, or working on writing a sentence on their own.

Materials for Step 9
- Oatmeal

FOCUS CHART

What are the parts of a redworm?

Worms have a head end and a tail. Worms have segments and a mouth.

ELA CONNECTION

These suggested strategies address the Common Core State Standards for ELA.

W 5: Strengthen writing.

SL 1: Participate in collaborative conversations.

Animals Two by Two Module—FOSS Next Generation

INVESTIGATION 3 – Big and Little Worms

No. 21—Teacher Master

No. 22—Teacher Master

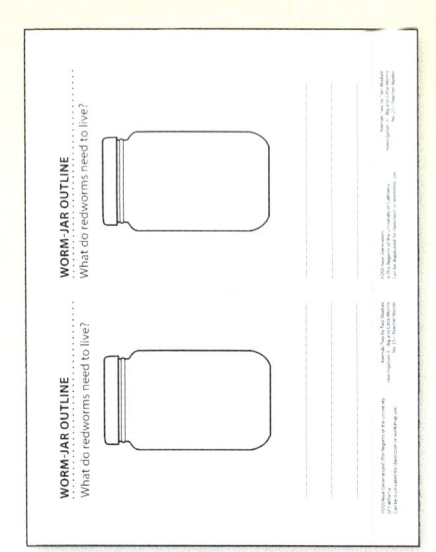

No. 23—Teacher Master

MATERIALS for
Part 2: Redworm Behavior

For each student at the center
- 1 Plastic cup
- 2 *Worm-Jar Outline* ★

For the class
- 1 Redworm terrarium (from Part 1)
- 6 Containers, 1/2 L
- • Water ★
- 1 Pitcher or water container ★
- 1 Plastic cup
- • Paper towels ★
- 1 Spray mister
- 1 Tray, white plastic
- 10 Newspaper sheets, single pages (no glossy pages) ★
- • Small pieces of apple ★
- • Small pieces of carrot ★
- • Small pieces of lettuce ★
- • Small, dry leaves, crumbled or torn ★
- • Flower petals or small weed plants ★
- • Grass seeds
- 10 Craft sticks
- 10 Plastic spoons
- 5 Jars, plastic, with lids with holes, 2 L (64 oz.)
- 8 Objects to use as barriers (See Step 3 of Getting Ready.) ★
- 5 Construction paper sheets, 30 × 46 cm (12" × 18") or file folders (See Step 5 of Getting Ready.) ★
- 5 Index cards (optional) ★
- • Packing tape, clear ★
- • Paper clips, regular and jumbo ★
- ❑ 1 Teacher master 21, *Center Instructions—Redworm Behavior A*
- ❑ 1 Teacher master 22, *Center Instructions—Redworm Behavior B*
- ❑ 1 Teacher master 23, *Worm-Jar Outline*

For assessment
- • *Assessment Checklists* 1 and 2

★ Supplied by the teacher.   ❑ Use the duplication master to make copies.

Full Option Science System

Part 2: Redworm Behavior

GETTING READY for
Part 2: Redworm Behavior

1. **Schedule the investigation**
   This part requires 15–20 minutes at the center for each group of six to ten students. Plan 5 minutes for an introduction and 15 minutes for students to write and draw in their notebooks.

2. **Preview Part 2**
   Students focus on the movement and behavior of redworms. They notice how the worm's body contracts and stretches to move forward. They observe the worm to see if it can move in other directions. They try blocking the worm's path to see what it does. Students consider what redworms need to live and set up a worm-jar habitat. Students observe how redworms change leftover food and other plant materials into soil. The focus question is **What do redworms need to live?**

3. **Gather objects for barriers**
   Gather at least eight small objects to use as barriers to block a worm's path. The objects should be waterproof and colorfast. Blocks, centicubes, pencils, and paper clips (regular and jumbo) are suitable choices. Put them in a 1/2 L container to use at the center. Keep the set together for Part 3.

4. **Gather materials for worm jars**
   At the end of the center observations, the group of students at the center will assemble a worm compost jar. Gather materials for students to add to the jar: sheets of newspaper (no glossy-colored sheets), an apple, carrot, small pieces of lettuce, small dry leaves, flower petals, or the stem and leaves from a small weed plant. The apple and lettuce can be old and not suitable for eating.

   Place each kind of material in a 1/2 L container. Cut or tear the single sheets of newspaper into quarters (about 14 × 27 cm). Each student will need one quarter of a sheet of newspaper.

   Students will also add a pinch of grass seeds to each jar. The grass seeds are provided in the kit. Put some seeds in a plastic cup.

5. **Make construction paper sleeves**
   Each worm compost jar needs a paper sleeve to make it dark inside the jar. Prepare one sleeve for each jar. You can use standard file folders or construction paper for the sleeves. Use construction paper sheets that are 30 × 46 cm. The file folders will provide a stiffer, sturdier sleeve, and will hold up to numerous up-and-down

> **TEACHING NOTE**
> Plan one or two short observation sessions each week for students to monitor changes taking place in the worm jars throughout the module (see Step 17 of Guiding the Investigation). This will provide evidence of how animals change their environment.

> **TEACHING NOTE**
> Students love to see worms crawl through the holes when several centicubes are connected.

Cut a newspaper page into quarters.

Animals Two by Two Module—FOSS Next Generation

INVESTIGATION 3 – Big and Little Worms

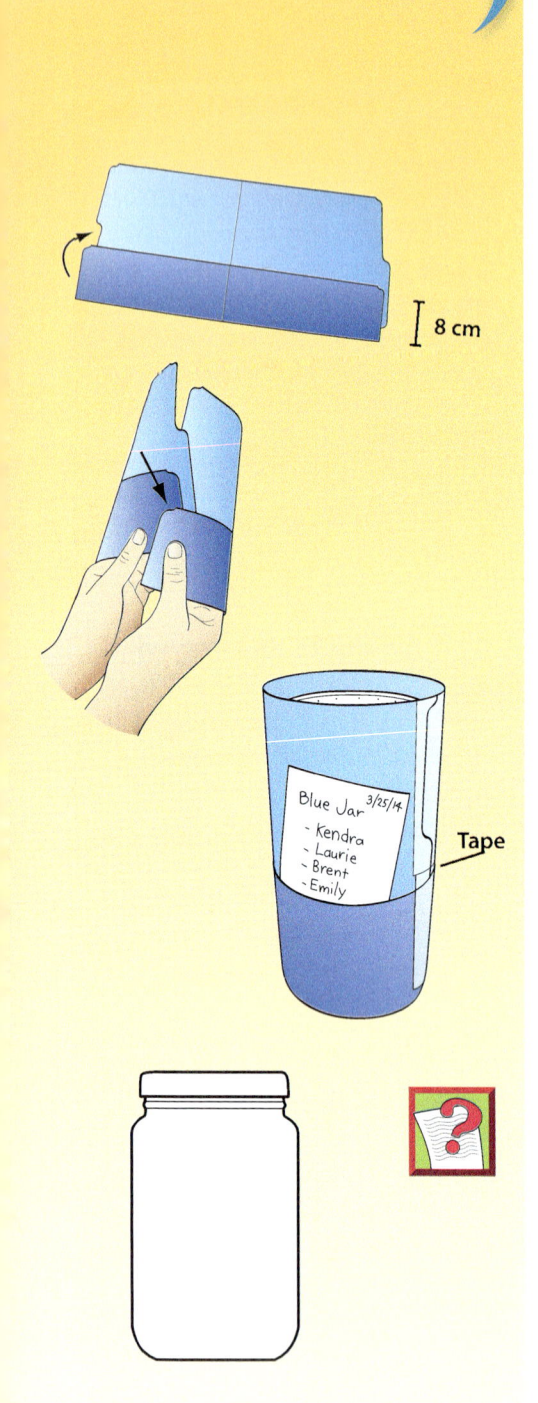

handlings to view the jar. Construction paper is thinner and will tear more easily. Both will work fine. Choose the one you have available.

Open a file folder and lay it flat. Fold up the bottom (long) edge about 8 cm and crease it. Find the plastic jar from the kit. Wrap the sleeve around the jar with the folded piece on the outside. Tuck one end of the fold into the other folded end. The sleeve should be snug around the jar, but still easy to slip off and on. Use two pieces of packing tape to secure the two vertical edges where the folded (8 cm) and the longer (23 cm) pieces come together. Repeat the process for the other file folders.

You can use index cards to label the jars with the group members' names. The short end of the card will slip nicely into the pocket of the sleeve. You can prepare the cards by labeling them "red jar," "blue jar," and so forth, to match the color of the sleeve, or leave the cards blank and allow students to add the date and names to the card after they build their worm jar.

6. **Set up the center**
Place the redworm terrarium in the center of the table. Put a plastic cup at each student's place. Keep a container of water and the containers of miscellaneous objects at hand. Keep a roll of paper towels handy to wipe up spills.

Gather the five jars with lids (the lids have small holes in them for air), the redworm food, grass seeds, plastic spoons, craft sticks, spray mister and paper sleeves, and place them at the center.

7. **Plan assessment for Part 2**
Plan to observe students as they assemble their redworm jars and record monitor changes over time.

What to Look For

- *Students describe observations about how redworms change their environment; students predict what changes will occur in the worm jar over time. (Constructing explanations; ESS2.E: Biogeology; cause and effect.)*

- *Students use a model of a worm jar to accurately show the relationship between the animal and the place where they live. Students see that animals live in places where their needs are met. (Developing and using models; ESS3.A: Natural resources; systems and system models.)*

Full Option Science System

Part 2: Redworm Behavior

GUIDING *the Investigation*
Part 2: **Redworm Behavior**

1. **Introduce the investigation**
   Call students to the rug. Ask them to share a few of their observations about a worm's structure.

   Tell them that today they are going to focus on how the earthworms move and how they behave, rather than how they look. Send six students to the center.

2. **Dig for redworms**
   When students arrive at the center, have them dig for worms to put in their cups. Allow 5 minutes for free observation of worms unguided by adult suggestions. Encourage and record students' questions.

   Pour a little bit of water into students' cups if they want to rinse the worms.

3. **Observe worm movement**
   When they're ready to move on, help students study movement and other behavior of the worms. Have students dip their fingers in a cup of water to wet the surface of the table. Have them place a worm on the table in the wet area and watch how it moves.

4. **Guide observations**
   After several minutes of observing worm movements, ask,

   ▶ *How do the worms move?*

   ▶ *Do they always* **move forward**? *Can you get them to move* **backward**? **Sideways**?

   ▶ *Does the worm change shape when it moves? Describe the shape.*

   ▶ *What questions do you have about redworms?*

5. **Block the worm's path**
   Show students the container of objects. Ask them how they could find out how the worms interact with objects. Explore any reasonable suggestions. If students do not suggest putting an object in front of the redworms, suggest it. Let students choose an object to put across the path the earthworm seems to be taking.

   ▶ *How does the redworm react to an object in its way?*

   ▶ *What do you think a snail would do if it came to the same object?*

   ▶ *How does the redworm get by the object?*

FOCUS QUESTION
What do redworms need to live?

EL NOTE
Model drawing a diagram of the redworm and have students help you label the parts.

Materials for Steps 2–5
- *Redworm terrarium*
- *Plastic cups*
- *Container of water*
- *Containers of objects*
- *Paper towels*

TEACHING NOTE
If students are reluctant to pick up the earthworms, let them observe until they decide they are ready to do it themselves.

SCIENCE AND ENGINEERING PRACTICES
Asking questions
Planning and carrying out investigations

Animals Two by Two Module—FOSS Next Generation

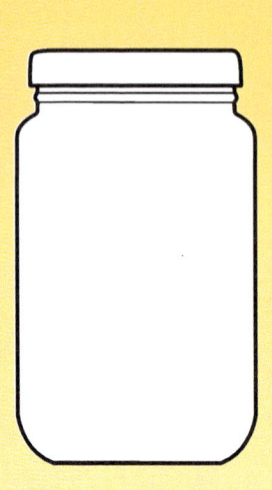

INVESTIGATION 3 – Big and Little Worms

EL NOTE

Start a chart to show the needs of worms. Add to it as students discover evidence of something that fulfills those needs.

Food	Water	Space/Air	Shelter
flower petals	spray bottle	Jar with holes in the lid	newspaper
weeds	lettuce		dried leaves
ryegrass seeds	apple		flower petals
lettuce, carrots			weeds

SCIENCE AND ENGINEERING PRACTICES

Planning and carrying out investigations

6. Introduce the worm jars
Ask,

➤ *What do worms need to live?*

Guide the discussion, and introduce food, water, air, and shelter if students don't mention them.

Show an empty plastic jar. Tell students that today they will make a home for worms in this jar. They will observe the worms over the next few weeks and see what happens.

7. Focus question: What do redworms need to live?
Write the focus question on the chart, and read it together.

➤ *What do redworms need to live?*

8. Display the materials
Show the materials that will go into the jar: newspaper, dried leaves, flower petals or weeds, ryegrass seeds, lettuce, apple slices. As students identify them, have them discuss what basic need they are providing (i.e., apple is food and water; leaves are shelter and food). If students don't mention newspaper as a food item, don't tell them. They will discover this on their own as they observe the jar over time.

9. Begin with newspaper
Give each student a one-quarter sheet of a newspaper page (no glossy sheets with color). Show them how to tear long strips of newspaper. Pass the jar around and have each student add their newspaper strips to the jar.

10. Add other items
Pass the containers with the other items to the students. Each student will add only one item to the jar. Let them add the item in the following amounts:

1   Small (5 cm) piece of lettuce torn into smaller pieces
1   Small piece of apple and carrot
1–2 Small dry leaves, crumbled or torn
1   flower petal or small weed plant
1   Scoop of grass seeds using a craft stick

11. Add worms and water
When the items have been added, pass around the containers with worms. Let each of six students add 4 worms to the jar (for a total of 24 worms per jar). If students don't want to handle them, they can use a plastic spoon to carefully transfer them to the jar.

Finally, students take turns giving one squirt of water to the jar.

Part 2: Redworm Behavior

12. Add lid and observe worms

Show the completed worm home and tell students they have lots of air in the jar but you are going to add a lid that has tiny holes in it for more air.

Pass around the jar one last time for each student to take a quick look. Tell students that the worms also need a dark place to live. Place the paper sleeve on the jar. Show them how the sleeve can be removed for observing the jar later.

13. Prepare for the next group

Have students help clean up the center. They should wash their hands prior to returning to their tables. Set up another jar and materials for the next group.

POSSIBLE BREAKPOINT

14. Observe worm jars as a class

Gather the class at the rug. Display the completed worm jars and ask,

➤ *What did we add to the worm jars?*

As students respond, write the items on the chart as a list on the left. On the right, draw an outline of a jar. Draw the items in the jar as they are called out.

When finished, summarize by asking,

➤ *Did we give the worms air, water, food, **shelter**? Where do you see it?*

Point to the appropriate places on the chart as students respond. Tell students that they gave the worms a dark place to live (the sleeve that covers each jar). Explain that the worms also need a cool place to live. Ask students where in the room they can keep the worm jars to keep the worms cool.

15. Review vocabulary

Add any new or important vocabulary to the class word wall. Let students be the guides—acknowledge the words they use and offer new vocabulary as needed.

16. Answer the focus question

Restate the focus question, and have the class read it aloud together.

➤ *What do redworms need to live?*

Tell students you have a sheet with the focus question and an outline of a jar on it. Distribute a copy of teacher master 23, *Worm-Jar Outline*, to each student and remind them how to glue it into their notebook.

CROSSCUTTING CONCEPTS

Systems and system models

▶ **NOTE**
Go to FOSSweb for *Teacher Resources* and look for the Crosscutting Concepts—Grade K chapter for details on how to engage young students with this concept.

backwards
forward
move
shelter
sideways

Animals Two by Two Module—FOSS Next Generation

INVESTIGATION 3 – Big and Little Worms

SCIENCE AND ENGINEERING PRACTICES
Developing and using models

▶ **NOTE**
Monitor the moisture level in the jar daily. It should be moist but not dripping wet with drops of water on the sides of the jar. If it looks or feels wet, unscrew the lid or take it off entirely to allow for evaporation. The worms will not crawl out.

TEACHING NOTE
Refer to the Sense-Making Discussions for Three-Dimensional Learning chapter in Teacher Resources on FOSSweb for more information about how to facilitate this with young students.

Ask students to answer the focus question in pictures and/or words. Tell them they can add drawings of items to the outline of the jar.

You might also want to give sentence frames to prompt their writing, such as We put ____ , ____ , ____ , and ____ in the jar for the worms. Or Worms need ____ to live. Or I think ____ will happen.

B R E A K P O I N T

17. Check the worm jars
Over the next few weeks, students will observe many changes in the jar. Early in this process, ask students to predict what the jar will look like in three weeks. Ask them to write or draw their prediction in their notebooks. In three weeks, they will return to their prediction and compare it to the actual results.

Here are some of the things students will see happen over several weeks.

- Seeds will sprout.
- Lettuce, apples, weeds, and newspaper will be eaten and reduced.
- The volume of material will become less as the worms mix up and eat the material.
- Worms move on the sides of the jar.
- Bits of soil (worm poop) accumulate on the sides and bottom of the jar

Set up a time once or twice a week for the students to observe the worm jars. Give them another sheet with an outline of a jar for them to record their observations. As students begin to notice the change in the worm jar environment, seize the opportunity to have students describe this phenomenon.

18. Have a sense-making discussion
Monitor the jars for wetness and food depletion. The jar should be kept damp. You shouldn't have to add any more water after the first day because of all the moisture in the food and the water squirts. Once a week, students can add a few leaves, lettuce, newspaper, or an item of their choosing. The jar should not become full of solid material. Leave plenty of air space for the air to circulate.

Occasionally at a center, you can carefully pour out the contents of the jar into the shallow tray. Students can dig through the contents to search for worms and other items. Give each student a cup and spoon to scoop up items, so they can observe them more closely.

Part 2: Redworm Behavior

Ask questions to guide their thinking about the phenomenon.

➤ *Where are the worms? What do they look like?*

➤ *Can you find any newspaper? Leaves? Seeds? Where did they go?* [Worms ate them.]

➤ *What is this dark stuff? Could it be **soil**?*

➤ *How did the soil get in the jar?* [Redworms eat the lettuce, apples, weeds, and newspaper and then "poop" out the soil.]

➤ *Go back and review your predictions of what the worm jar would be like after three weeks. Compare your predictions to what you observe today.*

➤ *What questions do you have about the worm jars?*

Discuss what the redworms contribute.

We observed that redworms eat plant material and their "poop" looks like soil. In fact, worm "poop" becomes part of the soil and makes it healthy. Redworms (or compost worms) are good for the soil, for plants that grow in the soil, and therefore, for humans who eat the plants.

WRAP-UP/WARM-UP

19. Share notebook entries

Conclude Part 2 or start Part 3 by having students share notebook entries. Ask students to open their science notebooks to the most recent entry. Read the focus question together.

➤ *What do redworms need to live?*

Ask students to pair up with a partner to

- share their answers to the focus question;
- explain their drawings.

Have students make another drawing of the worm jar and show how the environment has changed. Tell students to take turns explaining the changes to their partner. Provide a sentence frame such as,

Before _____ , now _____ .

I wonder _____ .

CROSSCUTTING CONCEPTS

Cause and effect

SCIENCE AND ENGINEERING PRACTICES

Asking questions

Analyzing and interpreting data

Constructing explanations

FOCUS CHART

What do redworms need to live?

Worms need air, water, food, and shelter.

Animals Two by Two Module—FOSS Next Generation

INVESTIGATION 3 – Big and Little Worms

MATERIALS for
Part 3: Comparing Redworms to Night Crawlers

For each student at the center
- 1 Plastic cup
- 1 *FOSS Science Resources: Animals Two by Two*
  - "Worms in Soil"

For the class
- Redworm terrarium (from Part 2, not the worm jars but the terrarium)
- Soil
- Water ★
- 1 Plastic cup
- 1 Set of objects (from Part 2) ★
- Night crawlers ★
- Paper towels ★
- ❑ 1 Teacher master 24, *Center Instructions—Comparing Redworms to Night Crawlers*
- 1 Big book, *FOSS Science Resources: Animals Two by Two*

For assessment
- *Assessment Checklists* 1, 2, and 3

★ Supplied by the teacher.   ❑ Use the duplication master to make copies.

No. 24—Teacher Master

182 Full Option Science System

Part 3: Comparing Redworms to Night Crawlers

GETTING READY for
Part 3: Comparing Redworms to Night Crawlers

1. **Schedule the investigation**
   This part requires 15–20 minutes at the center for each group of six to ten students. Plan about 5 minutes to introduce the center, and 10–15 minutes for students to write and draw in their notebooks. Plan on another 15-minute session for the reading.

2. **Preview Part 3**
   Students discover a new kind of worm in their terrarium—night crawlers. The new worms are much longer and fatter than the redworms. Students observe the two kinds of worms and compare the structures and behaviors of the two animals. The focus questions are **How are redworms and night crawlers different?** and **How are they the same?**

3. **Add night crawlers to the terrarium**
   Take the night crawlers you've been storing in the 1/2 L container in the refrigerator, and pour them and all the soil into the redworm terrarium. Don't tell students that you've added the night crawlers; they'll discover that when they begin digging.

4. **Plan to read** *Science Resources*: "Worms in Soil"
   Plan to read "Worms in Soil" during a reading period after completing this part.

5. **Plan assessment for Part 3**
   Plan to observe students as they compare redworms to night crawlers and to review their notebook entries.

   What to Look For

   - Students use firsthand observations and readings to compare different kinds of worms to develop explanations of worm structures and behaviors and answer questions about how worms meet their needs. (Constructing explanations; obtaining, evaluating, and communicating information; LS1.C: Organization for matter and energy flow organisms; patterns.)

   - Students support a claim with evidence about the importance of worms to soil. (Engaging in argument from evidence; ESS3.A: Natural resources; systems and system models.)

Animals Two by Two Module—FOSS Next Generation

183

INVESTIGATION 3 – Big and Little Worms

FOCUS QUESTIONS

How are redworms and night crawlers different?
How are they the same?

Materials for Step 2
- *Worm terrarium*
- *Plastic cups*
- *Cup of water*
- *Container of objects*
- *Paper towels*

SCIENCE AND ENGINEERING PRACTICES

Planning and carrying out investigations

Analyzing and interpreting data

Constructing explanations

CROSSCUTTING CONCEPTS

Patterns

GUIDING *the Investigation*
Part 3: Comparing Redworms to Night Crawlers

1. **Introduce the investigation**
   Tell students,

   You will have another chance to work with the worms. Look carefully to see if you can find a new kind of worm—worms that are different from the redworms.

2. **Dig for worms**
   When students arrive at the center, have them start digging for worms. They should remember to push the soil aside gently so they won't injure the worms. Expect some extra excitement when students begin to find the large night crawlers. There should be some redworms in the terrarium so students can compare the two kinds of worms.

   If students ask for water, put a small amount in their cups (just enough to cover the bottom of the cup) so students can wash the worms. Worms can be kept in the cup, held in students' moist hands, or monitored on a wet table.

3. **Compare worms**
   After several minutes of observation, ask students to compare the worms they have found.

   ▶ *Do all the worms look the same? How are they different?* [Size and breadth are the most obvious differences.]

   ▶ *Do you think they have the same number of segments?*

   ▶ *Do they move the same way?*

   ▶ *Are their bodies shaped the same way?* [The night crawlers are a bit flatter near the tail.]

   ▶ *Do you think they are all the same kind of worms? Why or why not?*

   ▶ *Can you feel any stiff hairs (**bristles**) on the large worms?*

4. **Name the two kinds of worms**
   Ask each student to put his or her worms in two groups. Students can work with a partner, if they choose, and combine their worms. Tell them the names of the worms.

   We have found two kinds of worms that live on land. The small worms are redworms, and the large worms are **night crawlers**.

   ▶ *Can you guess why the worms might have been given those names?*

   Let students discuss their ideas for a few minutes.

184   Full Option Science System

Part 3: Comparing Redworms to Night Crawlers

5. **Continue the investigation**
   Give students some time for free exploration. Put the container of objects on the table so students can use them. They can continue to work with the worms until interest wanes or it is time for the next group to come to the center. Ask students if they have any questions about the worms that they might investigate. Save these questions to discuss with the group in Step 14.

6. **Review vocabulary**
   As students offer their observations, add any new or important vocabulary to the class word wall. Let students be the guides—acknowledge the words they use and offer new vocabulary as needed.

7. **Have a sense-making discussion**
   Sense-making discussions can be conducted in small groups as part of center time or as a whole class once all students have completed the center activity and before answering the focus question.

   To summarize the comparisons that students made, draw a T-table to list differences. Ask students to help you come up with the words. Then generate a similar list for the ways that the worms are the same.

8. **Focus questions: How are redworms and night crawlers different? How are they the same?**
   Write the focus questions on the chart as you read them aloud.

   ➤ *How are redworms and night crawlers different? How are they the same?*

   Tell students that you have a strip of paper with the focus questions written on it. Remind them how to glue the strip into the notebook before they answer the question. When they return to their tables, they should answer the focus questions in their notebooks with pictures and words.

9. **Prepare the center for the next group**
   Have students return all their worms to the terrarium. Cover up the worms with some soil. Collect and clean out the cups (be careful not to wash soil down the sink). Have students wash their hands. Wipe off the table, and you're ready for the next group.

   NOTE: Night crawlers will only last a few days in the terrarium because they need cooler temperatures than redworms. If students will observe the night crawlers again, place them in the refrigerator where they will live for a month or so.

SCIENCE AND ENGINEERING PRACTICES
Asking questions

bristle
night crawler

FOCUS CHART

How are redworms and night crawlers different?

redworms	night crawlers
small	big
red-brown	brown
smooth	rough on belly

FOCUS CHART

How are redworms and night crawlers the same?

Both have segments.

Both have a head end and a tail end.

Both need water and food.

Both wiggle forward.

TEACHING NOTE

See the **Home/School Connection** for Investigation 3 at the end of the Interdisciplinary Extensions section. This is a good time to send it home with students.

INVESTIGATION 3 – Big and Little Worms

Worms in Soil

Worms are animals. Where do you find worms? These worms live in the soil. They get water and food from the soil.

37

ELA CONNECTION

These suggested strategies address the Common Core State Standards for ELA.

RI 1: Ask and answer questions about key details.

RI 2: Identify main topic and retell key details.

RI 3: Describe the connection between two ideas.

RI 10: Actively engage in group reading activities with purpose and understanding.

W 8: Gather information to answer a question.

SL 4: Describe with details.

CROSSCUTTING CONCEPTS

Patterns
Structure and function

READING in Science Resources

10. Read "Worms in Soil"

"Worms in Soil" extends students' learning about the role of worms in gardens and compost piles. Students observe where worms live, and learn what they eat. Not all worms look the same. Some even live in water. The reading provides an opportunity to compare different worms to those they observed in the classroom.

Show students the photograph on the first page and ask them what they think the reading will be about. Read the title together and ask students to share with a partner what they know about worms and soil and what questions they have.

Call on a few volunteers to share their questions and record them on chart paper. Tell students that this article will give them more information about different kinds of worms.

Refer to the content grid and add a row for earthworms. Tell students to listen and look for information about worm parts, what worms do, where worms live, and what worms need to live.

Read the article aloud from the big book, using strategies that will be most effective for your class. Pause to discuss key concepts in the reading, to compare photos, and to respond to questions.

11. Discuss the reading

Fill in the content grid using these questions as a guide.

➤ Where do worms live?
➤ Where are the segments on a worm?
➤ What do worms eat?
➤ How do worms make soil?
➤ Which worms look like the ones we have in the classroom?

Focus on "what they need to live" and ask students if they notice any patterns in what fish, birds, snails, and worms need to live.

186

Full Option Science System

Part 3: Comparing Redworms to Night Crawlers

12. Share information about photos

Tell students you have more information to share about these animals that might answer their questions. Flip through the pages a second time sharing some of the interesting features of the animals.

Page 37. Redworms. Redworms, or compost worms, live near the surface of the soil. These worms are probably on their way back down as they prefer darkness to light. Worms do not have noses or eyes. They sense the world and breathe through the skin of their bodies.

Page 38. Night crawlers. This larger earthworm makes deep burrows in the soil. Earthworms crawl out at night in search of leaf debris to eat. In 1 acre of soil there can be a million earthworms, eating up to 40 tons of leaves, stems, and dead roots a year.

Page 41. Ask students what they notice about the worms and the soil in this photograph. Read the text and model how you would try to figure out the meaning of the word *burrow*. For example,

The text says, "Some worms burrow deep in the soil." I see that worms are in little tunnels in the soil and I know that deep means far below. I think burrow means to dig. I'll check in the glossary to make sure.

Ask students to explain how burrowing in the soil changes the soil the worms are living in. [Burrowing mixes up the soil. Worm burrows make spaces for air to travel in and out of the soil.]

Page 43. Redworms. These close-ups of the worms show the "rings" or segments of their body. You can also see the main blood vessel and digestive tract. They run the length of the body. Worms' blood has hemoglobin, the same iron-rich protein in human blood that gives it the red color when full of oxygen.

Page 44. Night crawler. It's easy to spot the raised band or clitellum on a night crawler. After mating, an egg case will develop inside the clitellum. Later, the egg case will be deposited in the soil. Between one and five worms will emerge from the case.

Page 46. Leech (bottom). A leech is a type of worm that lives in water. This one is from the Danube river delta in Europe. Leeches are born with 32 segments and never have any more. This leech will hibernate in the mud. Size: 20 cm.

Page 47. Earthworm from Rwanda (bottom left). This earthworm is 25 cm long. That's long but there are earthworms in Australia and South Africa that grow to 3 meters (over 9 feet). Earthworms grow more segments as they get older and bigger.

SCIENCE AND ENGINEERING PRACTICES

Obtaining, evaluating, and communicating information

ELA CONNECTION

This suggested strategy addresses the Common Core State Standards for ELA.

RI 4: Ask and answer questions about unknown words.

CROSSCUTTING CONCEPTS

Patterns
Structure and function

INVESTIGATION 3 – Big and Little Worms

ELA CONNECTION

These suggested strategies address the Common Core State Standards for ELA.

RI 7: Describe the relationship between illustrations and the text.

RI 9: Identify similarities in and differences between two texts on the same topic.

CROSSCUTTING CONCEPTS

Cause and effect

13. Focus on photographs

Have students think about and discuss how they are able to gather information from the photographs in the article. Ask,

➤ *How do the photographs help you understand what the author is trying to say?*

Students should notice that the photographs show a magnified view to help them see the details of the worm. Focus on a few pages and discuss examples of how the images support what is said in the text. For example, page 40 shows an image of worms in the soil and an image next to it of a new plant. This supports the text, "The new soil helps young plants grow."

As an extension, students can compare information about worms presented in this article to how worms are described and illustrated in other texts. See FOSSweb for a list of recommended books.

Part 3: Comparing Redworms to Night Crawlers

WRAP-UP

14. Share notebook entries

Conclude Part 3 by having students talk with a partner about the guiding question for the investigation. They should use their notebooks as a reference. After sharing with a partner, ask for volunteers to talk about their ideas.

Read the question together.

➤ *What do animals such as worms need to live and grow?*

Have students think about what they have learned about worms by observing them, comparing them, and by reading and discussing the information in the article. Give them time to discuss the question.

Revisit the list of questions students generated to see if they can be answered. Choose one to discuss or provide a question that would lend itself to engaging in argument from evidence.

For example, you might use the following question.

➤ *Are worms good for soil?*

Discuss what students would have to do to show their claims were true. What evidence would they need to provide?

A reminder: Night crawlers will only last a few days in the terrarium because they need cooler temperatures than redworms. If students will observe the night crawlers again, place them in the refrigerator where they will live for a month or so.

DISCIPLINARY CORE IDEAS

LS1.A: Structure and function

LS1.C: Organization for matter and energy flow in organisms

ESS2.E: Biogeology

ESS3.A: Natural resources

SCIENCE AND ENGINEERING PRACTICES

Engaging in argument from evidence

TEACHING NOTE

Go to FOSSweb for Teacher Resources *and look for the* Science and Engineering Practices—Grade K *chapter for details on how to engage kindergartners with the practice of engaging in argument from evidence.*

ELA CONNECTION

This suggested strategy addresses the Common Core State Standards for ELA.

SL 1: Participate in collaborative conversations.

Animals Two by Two Module—FOSS Next Generation

INVESTIGATION 3 – Big and Little Worms

TEACHING NOTE

Refer to the teacher resources on FOSSweb for a list of appropriate trade books that relate to this module.

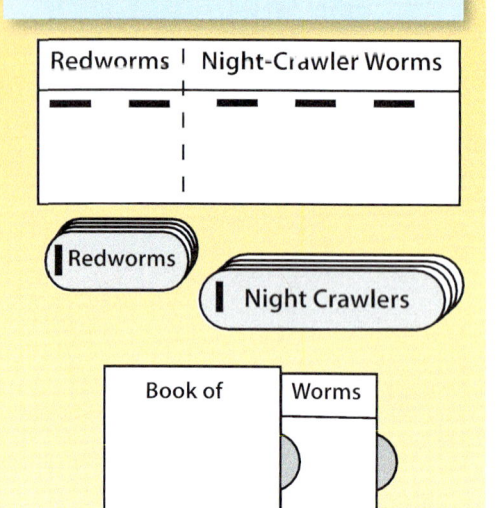

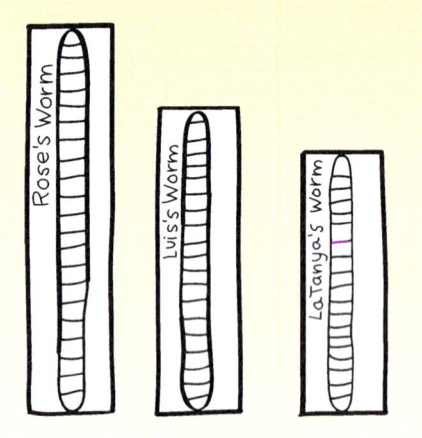

TEACHING NOTE

Review the online activities for students on FOSSweb for module-specific science extensions.

INTERDISCIPLINARY EXTENSIONS

Language Extension

- **Keep a classroom worm book**
  To make a worm journal, begin by folding a 30 × 46 cm piece of construction paper lengthwise, leaving a 5–8 cm margin along the top. Staple the top edge of the pocket. Fold the left side of the paper in one-third, forming a short and a long "worm tunnel." On the top margin of paper, label the short side "Redworms" and the long side "Night-Crawler Worms." Make two sets of worm book pages, each with matching construction-paper covers. The set of pages for the redworms should be no longer than 15 cm, and those for night crawlers no longer than 30 cm. Have half the class create a page to describe redworms and the other half to describe the night crawlers. Finally, staple each set together with its matching cover and store it in the "tunnels" of the book.

Math Extension

- **Compare the lengths of night crawlers**
  Cut strips of paper about 1 cm wide. Have students measure the length of a number of night crawlers by carefully laying each worm out on a strip of paper, marking the ends of the worm on the paper with a pencil, then cutting the paper to the length of the worm. Students can use nonstandard units such as jumbo paper clips to measure the length of the paper worm. (Jumbo paper clips are about 5 cm long and 1 cm wide.) Have them seriate the pieces of paper, from longest to shortest. Students might enjoy drawing the outlines of the worms on the strips and taking them home.

Science Extension

- **Look at worm paths**
  Have students dip their worms in water and let the wet worms crawl across dry construction paper. Have them compare the worm trails to the snail trails they observed in Investigation 2.

Interdisciplinary Extensions

Environmental Literacy Extensions

- **Take a schoolyard field trip**
  Look for worms after a rainstorm. Discuss what might cause them to come out of the ground. Have students dig for worms in the school garden, or put the class's worms into the garden to enrich the soil.

- **Set up a vermicompost system**
  The process of using worms and microorganisms to transform kitchen garbage (organic waste) into humus is known as vermicomposting. Work with the class to set up a large system. (See *Worms Eat My Garbage*, a teacher reference in the books section on FOSSweb.)

- **Invite a worm farmer to class**
  Almost every community is rich in resources providing worms for composting. Plan a visit to a worm farm to find out more about redworms. Or invite a vermicomposter to your class to explain how he or she cares for redworms.

Art Extension

- **Make worm tracks**
  Have students dip a piece of string into liquid tempera paint and then drag it across a piece of paper as a worm might crawl. They can continue dipping their strings and painting until their papers are covered with "worm tracks."

Home/School Connection

Families read questions and answers to students about the role of earthworms in the environment. Students cut out and match questions and answers.

Print or make copies of teacher master 25, *Home/School Connection* for Investigation 3, and give them to students to bring home anytime after Part 3.

As an additional home/school connection, ask students to talk to their parents about what they have learned about worms and to share experiences they have had with worms. Do they have a compost pile at home? Have they used worms for fishing?

TEACHING NOTE

Encourage students to use the Science and Engineering Careers Database on FOSSweb.

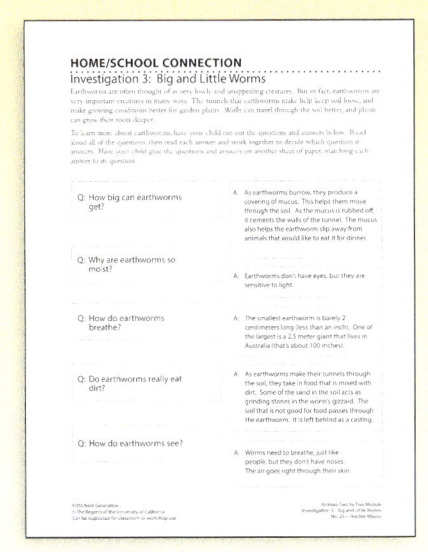

No. 25—Teacher Master

INVESTIGATION 3 – Big and Little Worms

INVESTIGATION 4 – Pill Bugs and Sow Bugs

Part 1
Isopod Observations............. 202

Part 2
Identifying Isopods............... 208

Part 3
Isopod Movement................ 215

Part 4
Animals Living Together....... 224

Guiding question for phenomenon:
What do animals such as isopods need to live and grow?

PURPOSE

Students have firsthand experiences with two closely related animals—pill bugs and sow bugs, two examples of the phenomenon called isopods. Through observation and discussion, students gather information about isopod structures and behaviors and how those characteristics relate to the needs of the animals.

Content

- Isopods are animals and have basic needs—water, air, food, and space with shelter.
- Different kinds of isopods have some structures and behaviors that are the same and some that are different.
- There is great diversity among isopods.
- Isopod behavior is influenced by conditions in the environment.

Practices

- Compare the structures and behaviors of two kinds of isopods, commonly known as pill bugs and sow bugs.
- Describe and communicate observations of several kinds of animals living together in a terrarium habitat.

Science and Engineering Practices

- Asking questions
- Planning and carrying out investigations
- Analyzing and interpreting data
- Constructing explanations
- Obtaining, evaluating, and communicating information

Disciplinary Core Ideas

LS1: How do organisms live, grow, respond to their environment, and reproduce?
LS1.A: Structure and function
LS1.C: Organization for matter and energy flow in organisms
ESS2: How and why is Earth constantly changing?
ESS2.E: Biogeology
ESS3: Earth and human activity
ESS3.A: Natural resources

Crosscutting Concepts

- Patterns
- Cause and effect
- Systems and system models
- Structure and function

FOSS Full Option Science System

INVESTIGATION 4 – Pill Bugs and Sow Bugs

	Investigation Summary	Time	Focus Question for Phenomenon, Practices
PART 1	**Isopod Observations** Students begin by investigating two kinds of isopods (sow bugs and pill bugs). They draw upon knowledge and experience gained from the previous activities to investigate the structures and behaviors of isopods.	**Introduction** 5–10 minutes **Center** 15–20 minutes **Notebook** 15 minutes	**What are isopods?** **Practices** Planning and carrying out investigations Analyzing and interpreting data
PART 2	**Identifying Isopods** Students compare the isopods and sort them into two groups, based on the different structures and behaviors they observe.	**Introduction** 5 minutes **Center** 15–20 minutes **Notebook** 15 minutes **Reading** 15 minutes	**How are pill bugs and sow bugs different? How are they the same?** **Practices** Asking questions Planning and carrying out investigations Analyzing and interpreting data Obtaining, evaluating, and communicating information
PART 3	**Isopod Movement** Students go to the schoolyard to find isopods. They discover where sow bugs and pill bugs live and observe their movement. In the classroom, students conduct isopod races as a way to focus observation on isopod movement.	**Introduction** 5 minutes **Outdoors** 15 minutes **Center** 15–20 minutes **Notebook** 15 minutes **Reading** 15 minutes	**How do isopods move?** **Practices** Planning and carrying out investigations Analyzing and interpreting data Constructing explanations Obtaining, evaluating, and communicating information
PART 4	**Animals Living Together** Students build a class terrarium to observe how several animals live together. They put the isopods and a few snails into the earthworm terrarium, then add objects from the natural environment to create an appropriate habitat for the animals.	**Introduction** 5 minutes **Center** 15–20 minutes **Notebook** 15 minutes **Reading** 35 minutes	**What do animals need to live?** **Practices** Planning and carrying out investigations Constructing explanations Obtaining, evaluating, and communicating information

At a Glance

Content Related to DCIs	Writing/Reading	Assessment
• Isopods are animals and have basic needs—water, air, food, and space with shelter. • Different kinds of isopods have some structures and behaviors that are the same and some that are different.	**Science Notebook Entry** Draw or write words to answer the focus question.	**Embedded Assessment** Teacher observation
• Different kinds of isopods have some structures and behaviors that are the same and some that are different. • There is great diversity among isopods.	**Science Notebook Entry** Draw or write words to answer the focus questions. **Science Resources Book** "Isopods"	**Embedded Assessment** Teacher observation
• Isopod behavior is influenced by conditions in the environment.	**Science Notebook Entry** Draw or write words to answer the focus question. **Science Resources Book** "Animals All around Us"	**Embedded Assessment** Teacher observation
• Isopods are animals and have basic needs—water, air, food, and space with shelter.	**Science Resources Book** "Living and Nonliving" **Book** *Animals Two by Two* **Online Activity** "Find the Parent"	**Embedded Assessment** Teacher observation **NGSS Performance Expectations addressed in this investigation** K-LS1-1 K-ESS2-2 K-ESS3-1

Animals Two By Two Module—FOSS Next Generation

INVESTIGATION 4 – Pill Bugs and Sow Bugs

BACKGROUND *for the Teacher*

What Are Isopods?

Iso is Greek for "similar" or "equal." *Pod* means "foot." Put them together and you have the **isopod**, an animal that has an equal number of feet (legs) on both sides and, more important, whose legs are all similar. Isopods have 14 legs that all function the same. This characteristic distinguishes them from related organisms, such as shrimps, crabs, and crayfish, which have legs that are modified to perform different functions, such as walking, feeding, and feeling.

Isopods are crustaceans, distant kin of the aquatic shellfish mentioned above. Like all crustaceans, isopods have a segmented outer **carapace** (seven overlapping plates) that provides protection from the environment and predators. Like their aquatic relatives, isopods get the oxygen they need through gill-like structures rather than through lungs like most terrestrial organisms. That is why isopods must keep moist at all times—if they dry, they die.

How Are Pill Bugs and Sow Bugs Different? How Are They the Same?

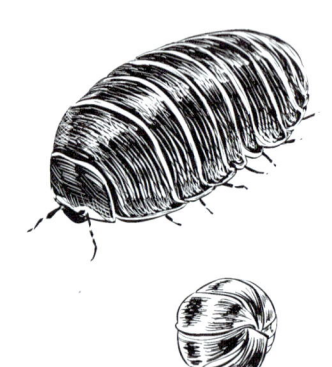

Two isopods are of interest as classroom organisms. The genus *Armadillidium* (arm•uh•duh•LID•e•um) is known casually as the **pill bug** or roly-poly because of its habit of rolling into a sphere when threatened or stressed. The pill bug **rolls up** in a **ball** to **protect** itself. The pill bug has a highly domed shape, short legs, and inconspicuous **antennae**. The tail end is **rounded**, not pointed. Pill bugs move slowly and have a difficult time righting themselves by **turning over** if they roll onto their backs on a smooth surface. They range from light brown to dark gray or black. Often they have whitish spots on their backs. The largest individuals of this kind of isopod can be 15 millimeter (mm) long, but most are 10 mm or less.

The second isopod used in classrooms, genus *Porcellio* (por•sel•E•oh), is commonly called the **sow bug** or wood louse. These names are potentially confusing because *Porcellio* don't show a particular affinity for swine, nor are they lice. They are **flatter** than pill bugs, have legs that extend a little bit beyond the edge of the shell, and have powerful antennae to sense their environment. They move rather quickly and use their long antennae and little spikelike tail projections to right themselves if they happen to roll onto their backs. Sow bugs come in a surprising array of colors, including tan, orange, purple, and blue, as well as the usual battleship gray. Their size is similar to that of the pill bug.

In the wild, isopods are not usually seen out and about. They are most often found in layers of leaf litter, under rocks or logs, or burrowed a

196 Full Option Science System

Background for the Teacher

short distance under the surface of the soil. The environment they seek is moist and dark, in or near dead and decomposing wood and other plant material. The former is their main source of food, accounting, perhaps, for their common name of wood louse. Isopods will eat fresh strawberries and carrots, making them a minor pest in the garden.

There are both male and female isopods, but only another isopod can reliably tell them apart. After mating, the female lays several dozen eggs, which she carries in a compact white package on her underside. This package is a specialized brood pouch, the marsupium, in which the eggs develop for 3–4 weeks before hatching. A few days after hatching, a swarm of fully formed, nearly invisible isopods strike out into the world. They soon grow to a size that can be seen by the unaided eye.

Like all crustaceans that wear a hard outer shell, isopods must shed their shells (molt) in order to grow. In the molting process, the shell is cast off, and the new soft shell underneath expands before hardening. Interestingly, the whole shell is not shed at once; first the rear (posterior) shell segments are shed, and 2–3 days later the front (anterior) segments fall off.

How Do Isopods Move?

Race them to find out. With their small, somewhat flattened bodies and 14 legs, isopods, particularly the sow bugs, can move very quickly. Sow bugs move forward, can turn quickly, and are constantly feeling the area in front with their antennae.

What Do Animals Need to Live?

Land animals all need variations of the same things—the right amount of water or **moisture**, food, air, and appropriate space with the right temperature and shelter. Isopods can live in just about any vessel, from a recycled margarine tub to a 50 liter (L) terrarium. If the container is smooth-sided, it doesn't have to be covered, because isopods can't climb smooth surfaces. A layer of soil covered with some dead leaves, twigs, and bark is great, but isopods will be comfortable with paper towels or newspaper laid on the soil—they do like to have some structure to crawl under. Sprinkle a little water on the soil and paper every day or two if it is open to the air. Keep a piece of fresh potato or carrot in the terrarium, and a piece of clear plastic wrap laid loosely over the lid, to ensure enough moisture for the isopods.

Before you put the isopods in the large earthworm terrarium, you will need to keep them for a few days in 1/2 L containers. When the 1/2 L containers are used to house isopods (or any other organism) for more than a day, poke a few holes in the lid. Keep the same conditions of moisture and food in the containers as described above.

"The environment they seek is moist and dark, in or near dead and decomposing wood and other plant material."

INVESTIGATION 4 – Pill Bugs and Sow Bugs

TEACHING CHILDREN *about* Isopods

Developing Disciplinary Core Ideas (DCI)

In this investigation, students are presented with two very similar organisms at the same time. Several differences between the two isopods are clearly observable, but students will have to concentrate to note them. Looking at two organisms simultaneously to discover similarities and differences is a challenging intellectual exercise. Students will have to look closely for subtle differences in structures, like curvature of carapace, length of legs, size of antennae, and shape of "tail," in order to discriminate between the two species. In their own minds, they will have to develop a system for comparing. Some students will be able to develop this structure unguided, but others will be able to make the comparisons only after the structure is provided. A leading question is sometimes effective: "Can any of the isopods close up into a ball? Is there a difference in how quickly some of the isopods can turn themselves over if they have been placed on their backs?" Narrowing and focusing allow students to make observations and draw conclusions.

Isopods are excellent classroom animals—they exhibit interesting behaviors; they are small but not tiny; they don't bite, smell, fly, or jump; and they are easy to care for. Isopods are very hardy animals, and students will enjoy handling them, but if their gills dry out, that's the end of them. So the most important thing to remember is to keep their terrarium moist at all times. You should put a time limit of 3–5 minutes for students to handle isopods outside of the moist environment.

It would be wonderful if students could study isopods next year and every year thereafter throughout their school careers. Each year they would see more, hear more, ask more, and learn more about these simple, unpretentious creatures. After students learned what isopods look like and how to tell one kind from another easily, they could ponder the isopod's environment—where it lives, what it needs, and what lives there with it. In due course, students would come to understand that no isopod lives forever and that death is part of life, but in spite of the mortality of individuals, life goes on. With any luck the colony of isopods with which students started their odyssey of life would still be healthy and happy. The only difference is that the 12th graders might be investigating the genetic code of the great, great, great, great, great, great, great, great, great, great grandchildren of the isopod they peered at for the first time in kindergarten.

NGSS Foundation Box for DCI

LS1.A: Structure and function
- All organisms have external parts. Different animals use their body parts in different ways to see, hear, grasp objects, protect themselves, move from place to place, and seek, find, and take in food, water, and air. Plants also have different parts (roots, stems, leaves, flowers, fruits) that help them survive and grow. (foundational)

LS1.C: Organization for matter and energy flow in organisms
- All animals need food in order to grow. They obtain their food from plants or from other animals. Plants need water and light to live and grow. (K-LS1-1)

ESS2.E: Biogeology
- Plants and animals can change their environment. (K-ESS2-2)

ESS3.A: Natural resources
- Living things need water, air, and resources from the land, and they live in places that have the things they need. Humans use natural resources for everything they do. (K-ESS3-1)

Teaching Children about Isopods

The activities and readings students experience in this investigation contribute to the disciplinary core ideas **LS1.A, Structure and function:** All organisms have external parts; **LS1.C, Organization for matter and energy flow in organisms:** All animals need food in order to grow; **ESS2.E, Biogeology:** Plants and animals can change their environment; and **ESS3.A, Natural resources:** Living things need water, air, and resources from the land, and they live in places that have the things they need.

Engaging in Science and Engineering Practices (SEP)

In this investigation, students engage in these practices.

- **Asking questions** about isopod habitats.
- **Planning and carrying out investigations** with isopods of several kinds to observe their structures and study their environmental needs; look at the relationship between the structures and how fast they move.
- **Analyzing and interpreting data** by describing observations of the isopods over time, recording information, using and sharing notebook entries, including writing and labeled pictures. Students use their firsthand observations and those of others in the classroom to describe the patterns they observe in isopod movement.
- **Constructing explanations** by making firsthand observations of isopods in the classrooms and those collected outdoors and using this as evidence to answer questions about the needs of animals, including food.
- **Obtaining, evaluating, and communicating information** about structures of isopods, their needs, and where they live.

NGSS Foundation Box for SEP

- **Ask questions** based on observations to find more information about the natural and/or designed world(s).
- **With guidance, plan and conduct an investigation** in collaboration with peers (for Grade K).
- **Make observations** (firsthand or from media) and/or measurements to collect data that can be used to make comparisons.
- **Make predictions** based on prior experiences.
- **Record information** (observations, thoughts, and ideas).
- **Use and share pictures, drawings,** and/or writings of observations.
- **Use observations** (firsthand or from media) to describe patterns in the natural world in order to answer scientific questions.
- **Compare predictions** (based on prior experiences) to what occurred (observable events).
- **Make observations** (firsthand or from media) to construct an evidence-based account for natural phenomena.
- **Read grade-appropriate text** and/or use media to obtain scientific and/or technical information to describe patterns in the natural world.
- **Communicate** information or solutions with others in oral and/or written forms using models and/or drawings that provide detail about scientific ideas.

INVESTIGATION 4 – Pill Bugs and Sow Bugs

NGSS Foundation Box for CC

- **Patterns:** Patterns in the natural and human designed world can be observed, used to describe phenomena, and used as evidence.
- **Cause and effect:** Events have causes that generate observable patterns. Simple text can be designed to gather evidence to support or refute student ideas about causes.
- **Systems and system models:** Objects and organisms can be described in terms of their parts. Systems in the natural and designed world have parts that work together.
- **Structure and function:** The shape and stability of structures of natural and designed objects are related to their function(s).

Exposing Crosscutting Concepts (CC)

In this investigation, the focus is on these crosscutting concepts.

- **Patterns.** Structures of different kinds of isopods are similar but they have differences in how they look, the appearance of the carapace (shape, color, pattern), and their behaviors.
- **Cause and effect.** Isopods can change their environment over time.
- **Systems and system models.** Isopods can be described in terms of their structures.
- **Structure and function.** The observable structures of isopods serve functions in survival.

Connections to the Nature of Science

This investigation provides connections to the nature of science.

- **Scientific investigations use a variety of methods.** Scientific investigations begin with a question. Scientists use different ways to study the world.
- **Scientific knowledge is based on empirical evidence.** Scientists look for patterns and order when making observations about the natural world.

New Word — Say it, See it, Hear it, Write it

Antenna
Ball
Carapace
Flat
Isopod
Jagged
Living
Moisture
Nonliving
Pill bug
Protect
Race
Roll up
Round
Section
Sow bug
Turn over

Teaching Children about Isopods

Conceptual Flow

Students continue to explore how animals of all kinds have various needs to live and grow. The fourth group of animals students study and care for live primarily on land. Isopods are the main phenomenon in this experience. The guiding question is what do animals such as isopods need to live and grow?

The **conceptual flow** for this first investigation starts with an introduction to isopods. Students observe several isopods, describe their parts (structures), and compare the variations in size and color. In Part 2, students learn that they have been observing two kinds of isopods—**pill bugs** and **sow bugs**. Students observe and describe the differences in their structures and behaviors.

In Part 3, students go outdoors in search of likely places to find isopods. They observe where isopods live and what they are doing. Back in the classroom, students focus on how the isopods move during races.

In Part 4, students assemble a terrarium with the land snails, earthworms, and isopods. They add other **living things** (plants and seeds) and some **nonliving things**—rocks, dead leaves, and bark. Students plan how to care for these animals living together in their classroom.

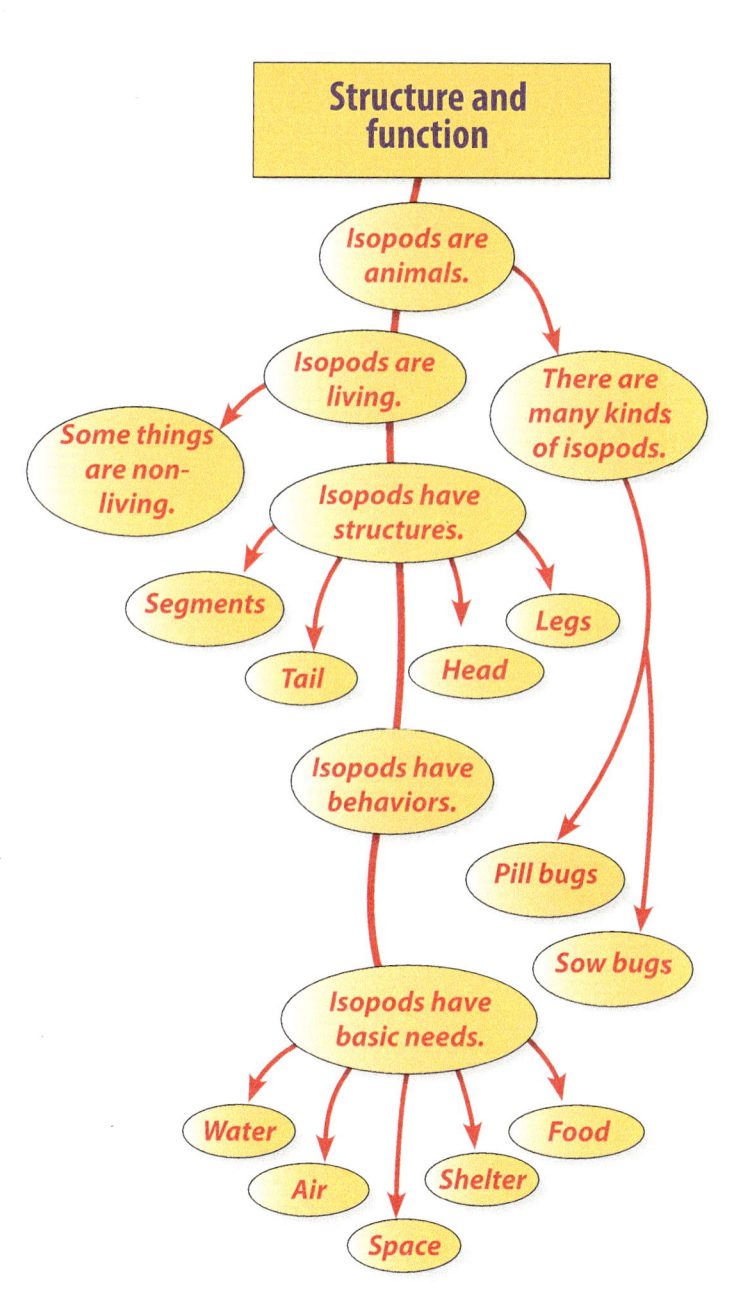

Structure and function
- Isopods are animals.
  - Isopods are living.
    - Some things are non-living.
  - There are many kinds of isopods.
  - Isopods have structures.
    - Segments
    - Tail
    - Head
    - Legs
  - Isopods have behaviors.
    - Pill bugs
    - Sow bugs
  - Isopods have basic needs.
    - Water
    - Air
    - Space
    - Shelter
    - Food

Animals Two By Two Module—FOSS Next Generation

INVESTIGATION 4 — Pill Bugs and Sow Bugs

MATERIALS for
Part 1: *Isopod Observations*

For each student at the center
- 1 Plastic cup
- 1 Paper towel square, 2 × 2 cm ★
- 1 Vial, 12 dram
- 1 Hand lens

For the class
- 3 Containers, 1/2 L
- 2 Container lids
- • Plastic wrap
- 25 Pill bugs (See Step 3 of Getting Ready.) ★
- 25 Sow bugs (See Step 3 of Getting Ready.) ★
- 2 Pieces of lettuce, potato or carrot ★
- 1 Set of objects to use as barriers (from Investigation 3, Part 2) ★
- • Paper towels ★
- 1 Scissors ★
- 5 Large bug boxes
- • Water ★
- ❏ 1 Teacher master 26, *Center Instructions—Isopod Observations*

For assessment
- ❏ • *Assessment Checklists* 1 and 2

★ Supplied by the teacher.   ❏ Use the duplication master to make copies.

No. 26—Teacher Master

202 Full Option Science System

Part 1: Isopod Observations

GETTING READY for
Part 1: *Isopod Observations*

1. **Schedule the investigation**
   Each group of six to ten students will need 15–20 minutes with the isopods. Plan an additional 5–10 minutes with the whole class to introduce the center and 15 minutes for students to write or draw in their notebooks at the end.

2. **Preview Part 1**
   Students begin by investigating two kinds of isopods (sow bugs and pill bugs). They draw upon knowledge and experience gained from the previous activities to investigate the structures and behaviors of isopods. The focus question is **What are isopods?**

3. **Obtain isopods**
   Look for isopods under stones, clay flowerpots, fallen tree branches, and plant litter on the ground, or in a compost bin. Or set out chunks of raw potato in your backyard overnight to attract them. Gather 25 pill bugs and 25 sow bugs. You can also order isopods from Delta Education. Be sure you ask for both pill bugs *(Armadillidium)* and sow bugs or wood lice *(Porcellio)*.

4. **Set up temporary isopod terrariums**
   Put wadded, moistened paper towels in two 1/2 L containers and add the isopods. Keep a piece of potato or carrot in each container for a backup moisture supply. Keep the lids on the containers.

5. **Obtain a set of objects**
   Gather a set of small miscellaneous objects from around the classroom. See Getting Ready for Part 2 in Investigation 3.

6. **Cut paper towels**
   Cut paper towel squares for each cup and sprinkle them with water.

▶ **NOTE**
To prepare for this investigation, view the teacher preparation video on FOSSweb.

Animals Two By Two Module—FOSS Next Generation

INVESTIGATION 4 – Pill Bugs and Sow Bugs

7. **Plan assessment for Part 1**
   Plan to listen and observe students as they work at the center and to review their notebook entries during or after class. Record your observations on *Assessment Checklists* 1 and 2.

   ### What to Look For

   - Students observe that animals (isopods) have external parts that help them to meet their needs. (LS1.A: Structure and function.)

   - Students make observations to collect data on isopod structures and compare these to different animals they have studied previously (Planning and carrying out investigations; patterns.)

Part 1: Isopod Observations

GUIDING the Investigation
Part 1: *Isopod Observations*

1. **Review the first three investigations**
   Call students to the rug. Ask them to recall the several kinds of animals they have recently investigated: goldfish and guppies, land and water snails, and redworms and night crawlers. For each pair of animals, have a student compare the structures and behaviors they observed. Use the content grid started in Investigation 1 as a reference.

2. **Introduce the isopods**
   Tell students,

   Today we will investigate some new animals called **isopods**. *We will find out as much as we can about them, what structures they have and how they use those parts, and what they need to live and grow.*

3. **Move to the center**
   Send six to ten students to the center and give each a plastic cup with a moist piece of paper towel in it. Pass the 1/2 L container of isopods around and ask students to pick out one or two isopods to put in their plastic cup. Students can also use the large bug boxes or just the lids to observe the isopods. Give students a few minutes to observe the isopods without a lot of adult guidance.

   It is OK for students to hold the isopods or put them on the table for 2–3 minutes.

4. **Observe structures of isopods**
   After several minutes, ask questions to help focus students' attention on the structures of the isopods.

   ➤ Who has seen this kind of animal before? Where did you see it?
   ➤ What do you call it?
   ➤ Which end is the head, and which is the tail? How do you know?
   ➤ How many legs does it have?
   ➤ Does it look the same on the top as on the bottom?
   ➤ How many **sections** do you think the **carapace** (hard outer covering) has?
   ➤ Where are the **antennae**?

   If a student mentions that the isopods can roll up into a **ball**, ask students to look at several more to see if all isopods can do that.

FOCUS QUESTION
What are isopods?

▶ **NOTE**
Isopods will be OK in a closed container with no air holes for a short time, up to a day. Poke holes in the lid with a nail or pushpin for longer stays.

EL NOTE
Give students a chance to repeat the word "isopod" out loud. You might emphasize each syllable. Encourage students to use the word in their explorations.

Materials for Step 3
- *Plastic cups*
- *Containers of isopods*
- *Container of objects*
- *Moist paper towels*
- *Bug boxes*

TEACHING NOTE
Students might notice that some isopods roll up; others don't. If students make this observation, great. Otherwise, don't focus on that behavior until Part 2.

Say it — See it — Hear it — Write it — **New Word**

Animals Two By Two Module—FOSS Next Generation

INVESTIGATION 4 – Pill Bugs and Sow Bugs

SCIENCE AND ENGINEERING PRACTICES
Planning and carrying out investigations

Materials for Step 7
- Vials
- Hand lenses

antenna
ball
carapace
isopod
section

EL NOTE
During the vocabulary review, draw and label a diagram of an isopod on chart paper for students to use as a reference.

5. **Investigate isopod movement and behavior**
Bring out the set of objects (blocks, centicubes, pencils, paper clips). Tell students they can use any of the items, one at a time, to investigate the isopods on the table top (out of the cup). Invite students to talk with a partner about which of the items they would like to use and why.

Tell students that when they are finished using something, they should return it to the container for others to use and then choose the next item they want to use. Let the investigations begin.

6. **Share investigations**
Ask students to share with others at the center what they tried and what they found out. Give them another few minutes to try new ideas they heard from other students.

7. **Introduce using the hand lens and vial**
Distribute a hand lens and vial to each student. Show them how to look at an isopod through a hand lens by placing the lens right on top of the open vial. Ask them to describe what they see. This is a homemade bug box. Have students put other isopods in bug boxes and compare how the two tools help them make observations of the animals.

8. **Review vocabulary**
As students offer their observations, add any new or important vocabulary to the class word wall. Let students be the guides—acknowledge the words they use and offer new vocabulary as needed.

9. **Focus question: What are isopods?**
Write the focus question on the chart as you read it aloud.

▶ What are isopods?

Tell students that you have a strip of paper with the focus question written on it. Describe how to glue the strip into the notebook before they answer the question. Ask them to return to their tables and work in their notebooks. Have them dictate a sentence for the bottom of the sheet or direct them to choose words to add from the class word wall.

Part 1: Isopod Observations

10. **Model responses to the focus question (optional)**
    Depending on students' experiences with notebooks, you can let them work on their own, or you can model making a notebook entry, using the focus chart.

11. **Prepare the center for the next group**
    Have students return all the objects to the container and all the isopods to their 1/2 L containers. Take the set of objects off the table until they are needed by the next group.

12. **Compare how animals move**
    Gather students at the rug. Ask students to talk with a partner about what structures isopods use for movement. Students should listen to their partner and then let them know if they agree or disagree with what their partner said. Ask them to do this for the other animals they have studied (fish, birds, snails, and worms.)

    Draw a graphic like the example shown and ask students to provide you with the information.

    Call on a volunteer to summarize how one animal moves. Call on a second volunteer to summarize how the next animal moves, and so on, until you finish by talking about how humans move.

WRAP-UP/WARM-UP

13. **Share notebook entries**
    Conclude Part 1 or start Part 2 by having students share notebook entries. Ask students to open their science notebooks to the most recent entry. Read the focus question together.

    ➤ *What are isopods?*

    Ask students to pair up with a partner to
    - share their answers to the focus question;
    - explain their drawings.

    Ask students to think about and discuss what they think the isopod parts are used for. Ask how the isopod parts are the same and different than those of a fish, a snail, or a worm.

SCIENCE AND ENGINEERING PRACTICES

Analyzing and interpreting data

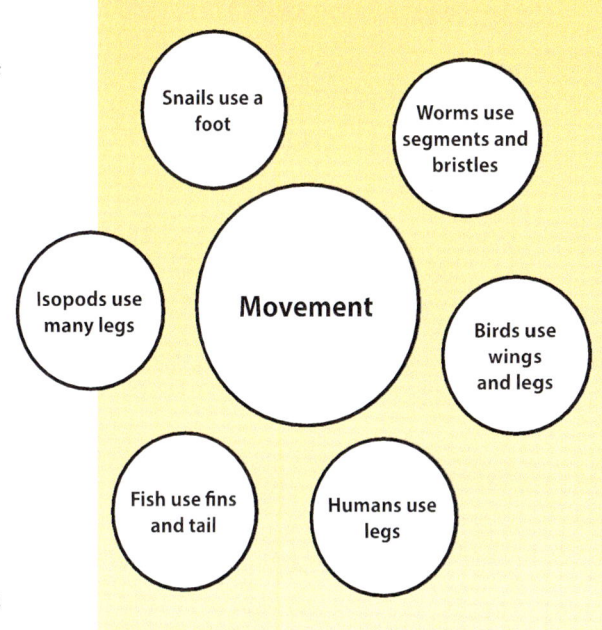

Movement
- Snails use a foot
- Worms use segments and bristles
- Isopods use many legs
- Birds use wings and legs
- Fish use fins and tail
- Humans use legs

FOCUS CHART

What are isopods?

Isopods are small animals with lots of legs. They have antennae. They have a hard carapace covering their body.

CROSSCUTTING CONCEPTS

Systems and system models

Animals Two By Two Module—FOSS Next Generation

INVESTIGATION 4 – Pill Bugs and Sow Bugs

MATERIALS for
Part 2: Identifying Isopods

For each student at the center
- 3 Plastic cups
- 1 Cup lid
- 1 Vial, 12 dram (large)
- 1 Hand lens
- 1 *Isopod Sorting*
- 1 *FOSS Science Resources: Animals Two by Two*
  - "Isopods"

For the class
- Isopods in containers★
- Paper towels ★
- Water ★
- 1 Spray mister
- ❏ 1 Teacher master 27, *Isopod Sorting*
- ❏ 1 Teacher master 28, *Center Instructions—Identifying Isopods*
- 1 Big book, *FOSS Science Resources: Animals Two by Two*

For assessment
- *Assessment Checklists* 1 and 2

★ Supplied by the teacher. ❏ Use the duplication master to make copies.

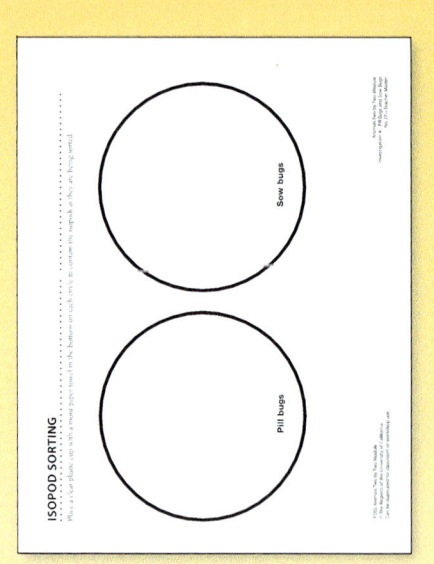

No. 27—Teacher Master

No. 28—Teacher Master

208 Full Option Science System

Part 2: Identifying Isopods

GETTING READY *for*
Part 2: *Identifying Isopods*

1. **Schedule the investigation**
   Each group of six to ten students will need 15–20 minutes at the center. Plan an additional 5 minutes to introduce the investigation to the whole class, 15 minutes for students to write or draw in their notebooks, and 15 minutes for reading.

2. **Preview Part 2**
   Students compare the isopods and sort them into two groups, based on the different structures and behaviors they observe. The focus questions are **How are pill bugs and sow bugs different?** and **How are they the same?**

3. **Set up isopod cups**
   Set up six to ten plastic cups (one for each student) with a piece of moist (but not soaking wet) paper towel in the bottom of each cup. Place two or three pill bugs and two or three sow bugs in each cup. Snap on lids securely. (See Background for the Teacher to identify the two kinds.)

4. **Plan to read *Science Resources*: "Isopods"**
   Plan to read "Isopods" during a reading period after completing the active investigation for this part.

5. **Plan assessment for Part 2**
   Plan to listen to and observe students as they work at the center and to review their notebook entries during or after class. Record your observations on *Assessment Checklists* 1 and 2.

   What to Look For

   - *Students use firsthand observations and readings to compare different kinds of isopods to answer questions about where isopods live. (Obtaining, evaluating, and communicating information; ESS3.A: Natural resources; patterns.)*

Animals Two By Two Module—FOSS Next Generation

INVESTIGATION 4 – Pill Bugs and Sow Bugs

FOCUS QUESTION

How are pill bugs and sow bugs different? How are they the same?

Material for Step 2
- Plastic cups
- **Isopod Sorting** sheets
- Cups of isopods
- Moist paper towels
- Cup lids

EL NOTE

Show students a picture of a sow bug and a picture of a pill bug from the Science Resources book before they start sorting.

SCIENCE AND ENGINEERING PRACTICES

Planning and carrying out investigations

Say it • See it • Hear it • Write it — **New Word**

CROSSCUTTING CONCEPTS

Patterns
Structure and function

GUIDING *the Investigation*
Part 2: Identifying Isopods

1. **Introduce the investigation**
   Call students to the rug. Ask them to share a few of their observations from the previous session. Tell them,

   There are actually two different kinds of isopods in the containers. Today you should try to separate them into two groups.

2. **Send students to center**
   Send six to ten students to the center.

3. **Look for differences**
   Give each student two empty cups and one cup containing pill bugs and sow bugs. Challenge students to carefully observe the isopods first and see if they can find a way to sort them into two kinds. They can talk with a partner about their ideas about how the kinds are different.

   Once they have a plan, each student should put one kind in one cup and the other kind in the other cup. If necessary, ask questions while students work on the task.

   ➤ *Look at the carapace (hard outer covering). Are they the same or different?*

   ➤ *Look at the antennae. Are they the same or different?*

   ➤ *Can all the isopods roll up into a ball?*

   ➤ *Can they turn over on their own? (If they are upside down with their legs in the air, can they turn over by themselves?)*

4. **Introduce the sorting mat**
   Distribute an *Isopod Sorting* sheet to each student. Read the two names in the circles. Say,

   *Some are **pill bugs** and some are **sow bugs**.*

   Ask a volunteer to describe which isopod might be the pill bug and have another student comment or build on what the first student says. Then confirm the differences.

   *Pill bugs have a much **rounder** carapace than sow bugs. A sow bug's carapace **flattens** out a bit on the sides, and the edge is **jagged**.*

   *Pill bugs can **roll up** into a tight little ball to **protect** themselves and conserve **moisture**. Sow bugs can't roll up.*

   *Sow bugs can **turn over** from their backs more easily than pill bugs.*

   Ask each student to put their isopods into one cup and then place an empty cup in each labeled circle. Challenge students to put the pill bugs in one cup and the sow bugs in the other.

Full Option Science System

Part 2: Identifying Isopods

5. **Provide more observation time**
   After students have identified the two kinds of isopods, give them a few more minutes to observe without adult guidance. Make available the large vials and hand lenses for making close observations.

   ➤ *Do you notice anything else about the two kinds of animals that helps you tell which are pill bugs and which are sow bugs?* [Accept all answers whether or not you think they have merit.]

6. **Review vocabulary**
   As students offer their observations, add any new or important vocabulary to the class word wall. Let students be the guides—acknowledge the words they use and offer new vocabulary as needed.

7. **Prepare the center for the next group**
   Ask each student to tell you one way that pill bugs and sow bugs are different before students leave the center. Have students return the isopods to the starting cups, and you're ready for the next group. After the last group has finished, return all the isopods to the 1/2 L containers. Mist the containers if they seem dry.

8. **Have a sense-making discussion**
   After all the students have sorted the isopods at the center, bring the class together for a discussion. Use a compare and contrast graphic organizer on chart paper as appropriate for your students. Ask them to talk with a partner about the differences between pill and sow bugs. As the two-person conversations wrap up, ask students to listen carefully to the ideas of their classmates. Facilitate a class conversation comparing the two kinds of isopods.

   Once there is agreement, add the idea to the chart. Once several ideas have been added to the chart, move on to how isopods are the same, and add those ideas to the chart.

9. **Focus questions: How are pill bugs and sow bugs different? How are they the same?**
   Write the focus questions on the chart as you read them aloud.

   ➤ *How are pill bugs and sow bugs different? How are they the same?*

   Tell students that you have a strip of paper with the focus questions written on it. Describe how to glue the strip into the notebook before they answer the question. When they return to their tables, they should answer the focus questions in their notebooks.

---

flat
jagged
moisture
pill bug
protect
roll up
round
sow bug
turn over

SCIENCE AND ENGINEERING PRACTICES
Analyzing and interpreting data

INVESTIGATION 4 – Pill Bugs and Sow Bugs

Isopods are animals. Where do you find isopods? These isopods live on land. Look for them under leaves and rocks where it is moist.

48

ELA CONNECTION

These suggested strategies address the Common Core State Standards for ELA.

RI 1: Ask and answer questions about key details.

RI 2: Identify main topic and retell key details.

RI 4: Ask and answer questions about unknown words.

RI 10: Actively engage in group reading activities with purpose and understanding.

W 8: Recall information from experiences to answer a question.

READING in Science Resources

10. Read "Isopods"

This article provides students with more information about the isopods they have been studying in class. It also provides images of other kinds of isopods.

Point out the title and have students read it with you chorally. Explain that *iso* means equal or the same and *pod* means foot. Ask students to think-pair-share why these animals are called isopods. They might have noticed that isopods have 14 feet—all of which function in the same way. Refer students to the content grid and add a row for isopods. Review the key details about isopods that students should listen and look for in the article—their parts, what they do, what they eat, and where they live.

Read aloud the article in the big book, using the strategies that will be most effective for your class. Pause to discuss key points in the article, to compare the photos, and to respond to the questions.

11. Discuss the reading

Fill in the content grid using these questions as a guide and to further the discussion.

➤ *Where do isopods live?*

➤ *What do they eat? What kind of evidence would you look for to know that they ate something?*

➤ *Where are the antennae? What do they do?*

➤ *Which isopods look like sow bugs? Pill bugs?*

➤ *What else did you learn about isopods?*

➤ *What questions do you have about isopods?*

12. Share information about photos

Tell students you have more information to share about these animals that might answer their questions. Flip through the pages a second time sharing some of the interesting features of the animals.

Page 48. Sow bugs (inset), pill bugs (top). Sow bugs prefer moist, dark places. They are often gray, flat, and have a "fringe" on the edge of their body.

Pill bugs are often darker, rounder, and prefer the drier side of a moist place. They can be found in sidewalk cracks and even crawling up concrete walls.

212   Full Option Science System

Part 2: Identifying Isopods

Page 49. Ask students what they notice about the photograph. Read the text and then discuss how isopods change the environment they live in by turning plant material into soil and moving the soil around.

Page 51. Pill bug (top). Pill bugs have a rounded segmented body that allows them to retain moisture. They can also curl into a ball to protect themselves from predators. Pill bugs and sow bugs have a pocket on their bellies for carrying their young.

Page 53. Aquatic isopods. Some isopods live in the ocean on kelp and in rock crevices.

Page 54. Rock louse (bottom left). A rock louse lives in the spray zone between the ocean waves and the dry land. It also needs to keep its gills wet so it can breathe. Size: 2.5 cm.

13. Focus on photographs

Have students think about and discuss how they are able to gather information from the photographs in the article. Ask,

➤ *How do the photographs help you understand what the author is trying to say?*

Students should notice that the photographs show a magnified view to help them see the similarities and differences between the types of isopods. Point out the photographs on pages 50 and 52 and ask how the images of a worm and a fish help the reader answer the questions posed in the text.

As an extension, students can compare information about isopods presented in this article to how isopods are described and illustrated in other texts. See FOSSweb for a list of recommended books.

SCIENCE AND ENGINEERING PRACTICES

Asking questions

Obtaining, evaluating, and communicating information

CROSSCUTTING CONCEPTS

Patterns

ELA CONNECTION

These suggested strategies address the Common Core State Standards for ELA.

RI 7: Describe the relationship between illustrations and the text.

RI 9: Identify similarities in and differences between two texts on the same topic.

Animals Two By Two Module—FOSS Next Generation

INVESTIGATION 4 — Pill Bugs and Sow Bugs

FOCUS CHART

How are pill bugs and sow bugs the same?

Both have a hard carapace.

Both have 14 legs.

Both need moisture.

Both eat potato or carrot.

FOCUS CHART

How are pill bugs and sow bugs different?

Pill bugs	Sow bugs
rounder	flatter
roll up	turn over
darker	long antennae
	tail

ELA CONNECTION

These suggested strategies address the Common Core State Standards for ELA.

W 5: Strengthen writing.

SL 1: Participate in collaborative conversations.

WRAP-UP/WARM-UP

14. Share notebook entries

Conclude Part 2 or start Part 3 by having students share notebook entries. Ask students to open their science notebooks to the last entry. Read the focus questions together.

➤ *How are pill bugs and sow bugs different? How are they the same?*

Ask students to pair up with a partner to

- share their answers to the focus questions;
- explain their drawings.

Have students critique their work by sharing with a partner one thing they like about their entry and one thing they can do to make it better. Suggest students use *FOSS Science Resources* book to help them get a better look at the isopod structures. They can use the photographs to help them make new drawings of a sow bug and a pill bug and label the parts using the words from the word wall.

Part 3: Isopod Movement

MATERIALS *for*
Part 3: *Isopod Movement*

For each student
- 1   *FOSS Science Resources: Animals Two by Two*
  - "Animals All around Us"

For the class
- 16   Plastic cups
- 16   Cup lids
- 1   Container, 1/2 L
- 2   Containers of isopods
- 1   Race track (See Step 8 of Getting Ready.)
- •   Pieces of potato, lettuce, or carrot ★
- •   Paper towels ★
- •   Water ★
- 1   Clock or stopwatch ★
- •   Transparent tape ★
- ❏ 1   Teacher master 29, *Center Instructions—Isopod Races*
- 1   Big book, *FOSS Science Resources: Animals Two by Two*

For assessment
- •   *Assessment Checklists* 1 and 2

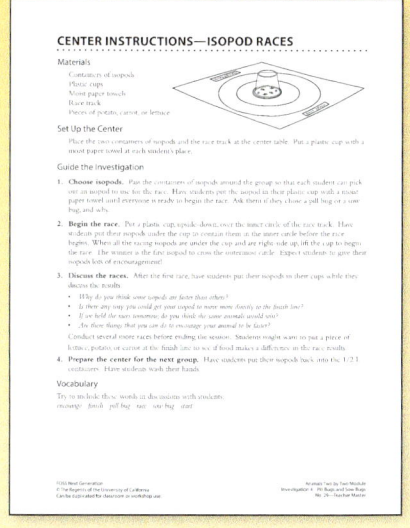

★ Supplied by the teacher.   ❏ Use the duplication master to make copies.

No. 29—Teacher Master

INVESTIGATION 4 – Pill Bugs and Sow Bugs

GETTING READY for
Part 3: *Isopod Movement*

1. **Schedule the investigation**
   The part begins with a 15-minute whole class excursion outdoors to search for isopods. Back in the classroom, you can work with the whole class at one time or at a center. Each group of six to ten students will need 15–20 minutes at the center. Plan an additional 15 minutes to introduce the investigation to the whole class and 15 minutes for students to draw and write in their notebooks. Plan one 15-minute session for reading.

2. **Preview Part 3**
   Students go to the schoolyard to find isopods. They discover where sow bugs and pill bugs live and observe their movement. In the classroom, students conduct isopod races as a way to focus observation on isopod movement. The focus question is **How do isopods move?**

3. **Select your outdoor site**
   Check around the school for isopods. Sow bugs are found under decomposing logs or in moist leaf litter. Search the base of bushes, hedges, and under planter pots. If you have a school garden, look there first. Pill bugs prefer slightly drier areas around sidewalks, concrete walls, and crevices. If it is very dry, isopods may be hard to find. Determine the outdoor boundaries for the activity and plan to describe them so students know where to search for isopods.

4. **Plan for additional help**
   If you conduct the outdoor search with the whole class, it is helpful to have additional helpers. Consider asking family members or older students to work with small groups of students to search for isopods. Extra hands and eyes will help the students to search safely.

5. **Plan for safety**
   Review outdoor safety rules and expectations before going outdoors. Hang up the FOSS *Outdoor Safety* poster and review it with students. If there are any areas or plants that are off limits, make sure to remind students and helpers about them.

6. **Check the site**
   Tour the outdoor site on the morning of the activity. Do a quick search for potentially distracting or unsightly items.

7. **Obtain a timer**
   This activity offers the opportunity for students to use a timer (clock or stopwatch) to time the races. Provide an appropriate timing device for students to use.

Full Option Science System

Part 3: Isopod Movement

8. **Prepare race tracks**
   Each race track comes in the kit as two separate laminated sheets, 28 × 43 centimeters (cm). Tape the two racetrack sheets together with transparent tape. There are enough sheets to make two race tracks.

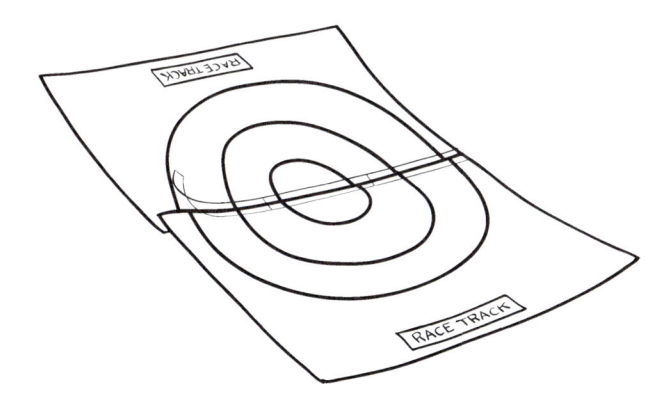

9. **Plan to read** *Science Resources*: **"Animals All around Us"**
   Plan to read "Animals All around Us" during a reading period after completing the active investigation for this part.

10. **Plan assessment for Part 3**
    Plan to listen and observe students as they work at the center and to review their notebook entries during or after class. Record your observations on *Assessment Checklists* 1 and 2.

    ### What to Look For

    - Students conduct isopod races and make observations to collect data on the movements of two different kinds of isopods. (Planning and carrying out investigations.)

    - Students interpret the results of the races and describe why they think some isopods are faster than others. (Constructing explanations; LS1.A: structure and function; cause and effect.)

Animals Two By Two Module—FOSS Next Generation

INVESTIGATION 4 – Pill Bugs and Sow Bugs

FOCUS QUESTION
How do isopods move?

Materials for Step 2
- Plastic cups
- Cup lids

▶ **NOTE**
Make sure to keep the isopods in enclosed cups out of direct sunlight. They can overheat quickly.

SCIENCE AND ENGINEERING PRACTICES
Planning and carrying out investigations

Materials for Step 4
- Container

GUIDING the Investigation
Part 3: Isopod Movement

1. **Discuss an isopod hunt**
   Tell students that today they will go outside to look for isopods. Ask them for ideas about where isopods might live and why. Encourage students to think about what type of area an isopod might prefer. Listen to their ideas and add suggestions of your own.

   Explain that they will work with a partner to find isopods. Each pair of students will have a plastic cup and lid to hold the animal when they find one. Every pair should find one or two. Before collecting the isopod in the cup, they should observe carefully what the animal is doing. They should remember where they found the isopod and what it was doing when they collected it.

2. **Go outdoors**
   Distribute a cup and lid to each pair of students and head outdoors to the selected outdoor area. Gather in the sharing circle to review safety rules and to describe the boundaries of the search area. Review the expectations for looking for isopods.

   - Treat all animals and plants with care.
   - If you are holding an isopod, hold it gently.
   - Stay with your partner and within the boundaries of the area.
   - When you locate an isopod, observe what it is doing. Then collect it and put it in your cup with the lid. Remember where you found each one.

3. **Search for isopods**
   Have students begin the search. Circulate to pairs as they observe and collect the animals.

   When each pair has collected one or two isopods, gather at the sharing circle. Ask a few pairs to share where they found the isopods and what they were doing.

4. **Return to class**
   Gather the materials and return to class with the collected isopods. You can either have students place all their isopods in one 1/2 L container or go on to the isopod races.

   Review and compare students' predictions about where they thought they would find isopods. Ask them to discuss what they found out about where isopods prefer to live.

Part 3: Isopod Movement

POSSIBLE BREAKPOINT

5. **Introduce the race track**

    Call students to the rug. Ask them to describe how the isopods moved. Students should say that isopods moved forward very fast, using their many legs.

    Show them the race track and ask them how they might use it with the isopods. Explain that today they will be having isopod **races**. Encourage students to ask questions and then offer the following.

    ➤ *Do you think the pill bugs or the sow bugs will move faster?*

    ➤ *If the animals started the race on their backs, which kind would win?*

    ➤ *If you had a race between a land snail and a pill bug, which do you think would win? Why do you think so?*

6. **Begin the races**

    Send six to ten students to the center, or work with the whole class at one time. Pass the containers of isopods around so each student can pick out an isopod to use for the race. Have students put the isopod in their plastic cup with a moist paper towel until everyone is ready to begin the race. Ask them if they chose a pill bug or a sow bug, and why.

    Have students take turns using a clock or timer to time how long a race takes.

    Put a plastic cup, upside-down, over the inner circle of the race track. Have students put their isopods under the cup before the race begins. When all the racing isopods are under the cup and right-side up, lift the cup to begin the race, and start the timer. Expect students to give their isopods lots of encouragement!

7. **Discuss the races**

    After the first race, have students put their isopods in their cups while they discuss the results.

    ➤ *How did the isopods move? Did they go in a straight line? Round and round motion? In a zig-zag motion?*

    Have students demonstrate each of those motions so you are sure they understand the movement.

    ➤ *How fast was your isopod?*

    ➤ *How many seconds or minutes did it take for your isopod to finish the race?*

    ➤ *Why do you think some isopods are faster than others?*

    ➤ *Do pill bugs or sow bugs move faster?*

SCIENCE AND ENGINEERING PRACTICES

Planning and carrying out investigations

Materials for Step 6
- *Race track*
- *Isopods in cups*
- *Plastic cup*
- *Clock or stopwatch*

CROSSCUTTING CONCEPTS

Systems and system models

Materials for Step 7
- *Carrot, lettuce, or potato pieces*

SCIENCE AND ENGINEERING PRACTICES

Analyzing and interpreting data

Constructing explanations

CROSSCUTTING CONCEPTS

Structure and function

INVESTIGATION 4 – Pill Bugs and Sow Bugs

CROSSCUTTING CONCEPTS

Cause and effect

race

➤ *Is there any way you could get your isopod to move more directly to the finish line?*

➤ *What might cause your animals to move faster?*

Conduct several more races before ending the session. Students might want to put a piece of lettuce, potato, or carrot at the finish line to see if food makes a difference in the results. Ask,

➤ *What do you think caused a certain isopod to move faster?*

8. **Review vocabulary**
   As students offer their observations, add any new or important vocabulary to the class word wall. Let students be the guides—acknowledge the words they use and offer new vocabulary as needed.

9. **Focus question: How do isopods move?**
   Write the focus question on the chart as you read it aloud.

   ➤ *How do isopods move?*

   Tell students that you have a strip of paper with the focus question written on it. Describe how to glue the strip into the notebook before they answer the question. When they return to their tables, they should answer the focus question in their notebooks with pictures, words, and numbers.

10. **Prepare the center for the next group**
    If you are working at a center, have students put their isopods back into the 1/2 L containers and wash their hands.

Part 3: Isopod Movement

READING in Science Resources

11. Read "Animals All around Us"

This reading extends students' understanding of similarities and differences in animals. Students read that some animals are covered with scales but that they're different from fish. Frogs have smooth, moist skin, but they are different from worms. Other animals have hair or fur. All animals have to fill basic needs to live. The reading provides an opportunity to compare new animals to the animals studied in the module.

Read the title of the article aloud and ask students if they know what animal is pictured on the first page of the reading. Ask students what other animals they think the article will be about.

Read the article aloud from the big book, using strategies that will be most effective for your class. Pause to discuss key ideas in the reading, to examine photos carefully, and to respond to questions.

12. Discuss the reading

Discuss the article, using these questions as a guide. If students have their own books, give them a few moments to locate the information in the article.

➤ *Which of these animals have legs? Which have scales?*
➤ *How are fish and lizards alike?*
➤ *How do some animals keep warm?*
➤ *What kind of skin do salamanders and worms have?*
➤ *Where do animals live?*
➤ *Is a bird an animal?*
➤ *Do you see any patterns in the structures that animals have?*
➤ *Do you see any patterns in the needs that animals have?*

If students have differing opinions, have them present their ideas and their reasoning. Encourage them to agree and disagree with each other and explain their thinking.

ELA CONNECTION

This suggested strategy addresses the Common Core State Standards for ELA.

RI 1: Ask and answer questions about key details.

SCIENCE AND ENGINEERING PRACTICES

Obtaining, evaluating, and communicating information

CROSSCUTTING CONCEPTS

Patterns

INVESTIGATION 4 – Pill Bugs and Sow Bugs

13. Share information about photos

Tell students you will walk through the article a second time and share some of the interesting features of the animals.

Page 56. Red fox (left), Gray wolf (right). Foxes and wolves are mammals. The red fox is an omnivore, eating small mammals, frogs, fruits, and vegetables. It uses its bushy tail to keep warm by wrapping it around its body.

The gray wolf eats deer, elk, and caribou. They also eat small prey like rabbits. Wolves are the ancestors of dogs.

Page 57. Common lizard (inset), Plains garter snake (bottom). Lizards and snakes are reptiles. The common lizard lives in Europe and Asia. It eats insects and bears live young. Body size: 12 cm.

This garter snake eats earthworms, slugs, and frogs. It lives in meadows near ponds and marshes. They bear an average of 15 live young. Size: up to 90 cm.

Page 59. Collared lizard (inset). This lizard lives throughout the western United States. It prefers pinyon-juniper, sagebrush, and desert grassland. The dark bands or collars on the neck give it its name. This is a female collared lizard because it is slightly green. The male is very green with dark spots on the throat. The female lays between one and thirteen eggs. Collared lizards eat insects, mice, and lizards.

Page 60. Bengal tiger (bottom). This tiger can live in bamboo forests in India. Their stripes help them to hide so they can surprise their prey at night. They eat buffalo, deer, and wild pigs.

Page 62. Red salamander (bottom). Salamanders are amphibians. They live part of their life in water and part on land. The red salamander takes cover during the day and is active at night. It eats insects and worms. Size: 10-15 cm.

Page 64. Bighorn sheep (left inset), Orca whale (right inset), White-tailed deer (bottom). These three animals are mammals. The bighorn sheep lives in rocky desert areas. They eat shrubs and grasses. The males have large horns to defend themselves from other males.

The orca whale has a distinctive white and black pattern. They live together in pods and eat fish and seals.

White-tailed deer get their name from the white patch on the underside of their tail. They live in the woods and are active in the early morning and late afternoon when they feed. Deer eat green plants, acorns, nuts, and buds from twigs in winter.

Part 3: Isopod Movement

If students have their own student books, have them choose one page from the article and take turns comparing the animals in the photograph with a partner. Encourage the use of language structures such as, "The _____ and the _____ are the similar because they both _____. They are different because the _____, but the _____."

WRAP-UP/WARM-UP

14. Share notebook entries

Conclude Part 3 or start Part 4 by having students share notebook entries. Ask students to open their science notebooks to the last entry. Read the focus question together.

➤ *How do isopods move?*

Ask students to pair up with a partner to

- share their answers to the focus question;
- explain their drawings.

ELA CONNECTION

This suggested strategy addresses the Common Core State Standards for ELA.

SL 2: Ask and answer questions about key details and request clarification.

FOCUS CHART

How do isopods move?

Isopods move quickly in a straight line, using their many legs. Sometimes they stop and hide.

Animals Two By Two Module—FOSS Next Generation

INVESTIGATION 4 – Pill Bugs and Sow Bugs

MATERIALS for
Part 4: Animals Living Together

For each student

- 1   *FOSS Science Resources: Animals Two by Two*
  - "Living and Nonliving"

For the class

- 1   Earthworm terrarium (from Investigation 3)
- 2   Containers of isopods
- 1   Land-snail terrarium (from Investigation 2)
- 1   *Animals Two by Two* book (See Step 7 of Getting Ready.)
- •   Ryegrass seeds
- •   Leaf litter ★
- 1   Small garden plant ★
- 1–2   Flat rocks ★
- •   Tree bark or stick ★
- •   Lettuce ★
- •   Pieces of potato ★
- ❏ 1   Teacher master 30, *Center Instructions—Animals Living Together*
- 1   Big book, *FOSS Science Resources: Animals Two by Two*
- 1   Projection system ★
- 1   Computer with Internet access ★

For assessment

- •   *Assessment Checklists* 1, 2, and 3

★ Supplied by the teacher.   ❏ Use the duplication master to make copies.

No. 30—Teacher Master

Full Option Science System

Part 4: Animals Living Together

GETTING READY for
Part 4: Animals Living Together

1. **Schedule the investigation**
   After a 5-minute introduction, each group of six to ten students will need 15–20 minutes at the center. Plan 15 minutes for students to draw and write in their notebooks. Plan an additional 15 minutes for a reading in *Science Resources* and a 20-minute final reading session for the book by Lawrence Lowery, also called *Animals Two by Two*.

2. **Preview Part 4**
   Students build a class terrarium to observe how several animals live together. They put the isopods and a few snails into the earthworm terrarium, then add objects from the natural environment to create an appropriate habitat for the animals. The focus question is **What do animals need to live?**

3. **Gather living and nonliving things**
   Gather the living and nonliving things (plant, ryegrass seed, leaf litter, flat rocks, tree bark or stick, lettuce, pieces of potato) needed for the terrarium.

4. **Plan for the end of the module**
   If you do not have to rotate the module to another teacher for a few weeks, build your combined terrarium in the established earthworm terrarium. If you need to pass the module on, find another clear plastic container to keep the terrarium going for as long as you like.

   Likewise, the fish will need to be transferred to another aquarium if the module is passed on to another teacher. Any snails that will not go into the terrarium can be passed to the next teacher who will use the module. If continued care of snails is not possible, the most humane way to end their lives is to freeze them. Snails should never be released into local environments if they were not collected there. Goldfish and guppies should not be released in the local environment. Redworms and isopods can be added to a compost bin.

5. **Plan to read *Science Resources*: "Living and Nonliving"**
   Plan to read "Living and Nonliving" during a reading period after completing the active investigation.

Animals Two By Two Module—FOSS Next Generation

INVESTIGATION 4 – Pill Bugs and Sow Bugs

6. **Plan for online activity**
   Preview the online activity "Find the Parent" and plan to use it toward the end of this part.

7. **Plan to read** *Animals Two by Two*
   Read through the book *Animals Two by Two* by Lawrence Lowery so you are familiar with the story. Plan to read that during a reading period at the end of the module.

8. **Plan assessment for Part 4**
   Plan to observe students as they modify the fish aquarium and model what happens.

   What to Look For

   - *Students work together to develop a terrarium for different land animals living together. Student see that animals live in places where their needs are met. (LS1.C: Organization for matter and energy flow in organisms; LS3.A: Natural resources; systems and system models.)*

   - *Students use examples to describe differences between living and nonliving things. (Constructing explanations.)*

Part 4: Animals Living Together

GUIDING the Investigation
Part 4: Animals Living Together

1. **Focus question: What do animals need to live?**
   Call the class to the rug. Introduce the focus question on the chart as you read it aloud.

   ➤ *What do animals need to live?*

   Tell students,

   Today we are going to set up a terrarium as a model to help us understand why the animals we have been studying live where they do. We are going to make a terrarium where the animals can live together. Each group will add some of the animals and some other objects to make a class terrarium. We will need to think about what the animals need to live together.

2. **Guide discussion**
   Call a small group to the center. Show students the earthworm terrarium. Tell them that this will be the home for the earthworms, the isopods, and a few land snails. Discuss what these animals will need to have a healthy and comfortable habitat.

   ➤ *What conditions do these animals need to live in the terrarium? (For example, wet or dry?)*

   ➤ *What kind of food will each of them need?*

   ➤ *What other things should we add to provide a suitable home for the animals?*

3. **Introduce nonliving things**
   Show students the nonliving things that you have for the terrarium—rocks. Tell them that these things can also be put into the terrarium.

   Explain to students that **living** means *alive* and **nonliving** means *not alive*. Animals and plants are living, but rocks are nonliving. Animals and plants are alive because they change and grow. All living things have basic needs and can have babies (offspring.)

4. **Assemble animals**
   Pass the isopod containers around and let each student catch an isopod and add it to the earthworm terrarium. As a group, have them add one snail. After they have added the living things, have them add a nonliving thing, such as a rock, or other object students think will make a nicer place for the animals to live.

FOCUS QUESTION
What do animals need to live?

SCIENCE AND ENGINEERING PRACTICES
Planning and carrying out investigations

Materials for Steps 2–5
- *Earthworm terrarium*
- *Containers of isopods*
- *Land-snail terrarium*
- *Leaf litter*
- *Small plants*
- *Flat rocks*
- *Tree bark*
- *Bits of lettuce and potato*
- *Grass seed*

New Word: Say it, See it, Hear it, Write it

Animals Two By Two Module—FOSS Next Generation

INVESTIGATION 4 – Pill Bugs and Sow Bugs

TEACHING NOTE

*If students have completed the **FOSS Trees and Weather Module**, they will know the basic needs of plants—water, light, air, nutrients from soil, and space. This is a good time to review those needs and compare them to the basic needs of animals.*

CROSSCUTTING CONCEPTS

Systems and system models

living
nonliving

DISCIPLINARY CORE IDEAS

LS1.A: Structure and function

LS1.C: Organization for matter and energy flow in organisms

ESS2.E: Biogeology

ESS3.A: Natural resources

TEACHING NOTE

See the Home/School Connection for Investigation 4 at the end of the Interdisciplinary Extensions section. This is a good time to send it home with students.

5. **Add plants**
   Tell students that you have other living things for them to add to the terrarium—plants. You can have them plant the small plant and some grass seeds in the terrarium.

6. **Discuss ongoing care**
   Ask students what they think they will need to do to keep the terrarium going for an extended period of time. Discuss the animals' and plants' needs, and set up a maintenance schedule for the class to follow.

   ➤ Where in the room should we place the terrarium?
   ➤ How much water will be needed?
   ➤ How much food should we put in the terrarium? How often should we feed the animals?

7. **Review vocabulary**
   As students offer their observations, add any new or important vocabulary to the class word wall. Let students be the guides—acknowledge the words they use and offer new vocabulary as needed.

8. **Answer the focus question**
   Read the focus question aloud with the class.

   ➤ What do animals need to live?

   Tell students that you have a strip of paper with the focus question written on it. Ask students to make a drawing in their notebooks of the terrarium and show how the terrarium meets the needs of that animal. Ask them to return to their tables and work in their notebooks.

   Ask each student to tell you one thing that animals need to live before they leave the center.

9. **Prepare the center for the next group**
   The class terrarium is a cumulative project; the next group will add isopods, a snail, and a few objects.

10. **Add other organisms**
    Over time, students might suggest adding other locally collected plants and animals to the terrarium. Before making any additions, discuss the needs of the new organisms as well as those already in the terrarium. Guide students in making appropriate decisions on what to add (or not add) to the terrarium.

Part 4: Animals Living Together

READING *in Science Resources*

11. Read "Living and Nonliving"

This final reading allows students to consider what things are living and what are nonliving.

Review with students what is living and nonliving in the terrarium. Tell students that this article will give them many more examples to discuss.

Read the article aloud from the big book, using strategies that will be most effective for your class. Pause to discuss key ideas in the reading, to examine photos carefully, and to respond to questions.

12. Discuss the reading

Discuss the article, using these questions as a guide. Turn to the appropriate pages in the big book, or have students use their own books to locate the information.

➤ *What is living in a garden?* [Plants, grass, flowers, bees.]

➤ *What are some nonliving things in a garden?* [Rocks, water, soil.]

➤ *What is another name for babies?* [Offspring.]

➤ *What animals lay eggs?* [Lizards, beetles, fish, birds.]

➤ *Name an animal that has their babies live.* [Mammals, such as dogs, cats, pigs, and humans.]

➤ *How are living and nonliving things different?* [Living things have basic needs; they grow and change; they can have offspring. Nonliving things don't have basic needs, don't grow, and can't have offspring.]

As an extension, provide images of other living and nonliving things and have students work with a partner to sort and explain why they think each is living or nonliving.

Living and Nonliving

Look at the garden.
What is **living** in the garden?

The garden has many plants.
Animals live here, too.

SCIENCE AND ENGINEERING PRACTICES

Constructing explanations

ELA CONNECTION

These suggested strategies address the Common Core State Standards for ELA.

RI 1: Ask and answer questions about key details.

RI 10: Actively engage in group reading activities with purpose and understanding.

L 5a: Sort objects into categories.

POSSIBLE BREAKPOINT

Animals Two By Two Module—FOSS Next Generation

INVESTIGATION 4 – Pill Bugs and Sow Bugs

Materials for Step 13
- *Animals Two by Two* trade book

SCIENCE AND ENGINEERING PRACTICES
Obtaining, evaluating, and communicating information

CROSSCUTTING CONCEPTS
Patterns

Systems and system models

ELA CONNECTION
These suggested strategies address the Common Core State Standards for ELA.

RI 2: Identify main topic and retell key details.

RI 6: Name and define the role of the author and illustrator.

RI 8: Identify the reasons an author gives to support points.

13. Read *Animals Two by Two*

Ask students how they can tell a pill bug from a sow bug. How can they tell a redworm from a night crawler, a goldfish from a guppy, and a land snail from a water snail? Explain that many animals look alike but are different. Brainstorm other animals that look alike but are really different.

Introduce the book *Animals Two by Two*. Ask students who the author is and what an author does. Ask who the illustrator is and what an illustrator does.

Read the book *Animals Two by Two* aloud, and ask students to listen for animals that are alike but different. Ask students questions and let them interject comments comparing the animals in the book and the animals they have been observing over the last few weeks.

14. Discuss the book

Tell students to think-pair-share what they think is the main idea of the book. Confirm that the main idea is that animals are similar but if we look closely and compare, we can see that they are also different. Discuss how the author and illustrator support this idea in the book. Encourage students to cite examples from the text.

They can use the illustrations in the book to retell the ways animals can be compared.

Use these questions to check for understanding and to further the discussion.

➤ What animals are alike but different?

➤ How are the animals different?

➤ What features do you look at when comparing two animals? [Shape, size, color, eyes, arms, legs, ears, toes, nose, etc.]

This is a good time to return to the animals that students have been observing in the classroom and to review how they are similar and different.

15. View online activity: "Find the Parent"

In small groups or as individuals, have students engage with the online activity "Find the Parent." This online activity gives students more opportunities to study animals of different kinds. The link to this activity for teachers is in the Resources by Investigation on FOSSweb.

Full Option Science System

Interdisciplinary Extensions

INTERDISCIPLINARY EXTENSIONS

Language Extension

- **Make a classroom isopod journal**
  Make one class journal for students' isopod observations. Make the journal in the image of an isopod. Round the corners on a stack of drawing paper and a black construction-paper cover, use brads for eyes, and cut seven little legs on each side of the cover. This journal stays with the isopod container so students can add their observations throughout the day.

Art Extension

- **Make a classroom mural**
  Create a mural in your classroom to show where isopods, worms, snails, and other terrestrial creatures can be found in nature. Enlist students in adding to a cross section of earth—underground, at the ground surface, and above.

Science Extension

- **Find out about other crustaceans**
  Isopods are crustaceans. Have students use the computer or books to collect information about other crustaceans—crabs, lobsters, crayfish, shrimp, and barnacles.

Environmental Literacy Extensions

- **Take a trip to the schoolyard**
  Have students go outdoors to the schoolyard and make a list of living and non-living things they observe. This could become another class mural by reproducing a model of the schoolyard habitat.

- **Improve schoolyard habitats**
  Take students to the schoolyard to survey possible animal habitats. Have them develop ideas for things to change in the schoolyard to improve the habitat for animals. Some ideas might be to design bird feeders and bird baths, grow plants that attract insects, or water some areas that are dry.

TEACHING NOTE

Refer to the teacher resources on FOSSweb for a list of appropriate trade books that relate to this module.

TEACHING NOTE

Encourage students to use the Science and Engineering Careers Database on FOSSweb.

TEACHING NOTE

Review the online activities for students on FOSSweb for module-specific science extensions.

Animals Two By Two Module—FOSS Next Generation

INVESTIGATION 4 — Pill Bugs and Sow Bugs

Home/School Connection

Students color the animals on the right-hand side of the page. Next they cut them out and glue them on the drawing where each animal lives. Students label one living and one nonliving thing in the picture. They share what they know about these animals with family members.

Print or make copies of teacher master 31, *Home/School Connection* for Investigation 4, and give them to students to bring home after Part 4.

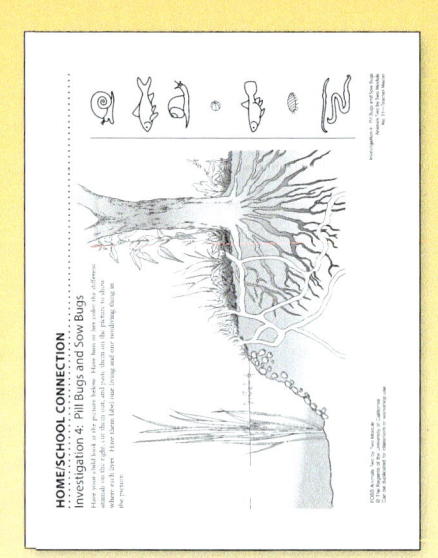

No. 31—Teacher Master

ANIMALS TWO BY TWO – *Assessment*

THE FOSS ASSESSMENT SYSTEM FOR KINDERGARTEN

"Assessment is like science. …To assess our students, we plan and conduct investigations about student learning and then analyze and interpret data to develop models of what students are thinking. These models allow us to predict the effect of additional teaching, addressing the patterns we notice in student understanding and misunderstanding. Assessment allows us to improve our teaching practice over time, spiraling upward" *(2016 Science Framework for California Public Schools, Kindergarten through Grade 12,* chapter 9, page 3*)*.

The FOSS K–8 assessment system is designed to assess students in cycles: short, medium, and long. The assessment tasks allow students to demonstrate their facility with three-dimensional understanding of science in both formative and summative assessments throughout a module.

FOSS assessment for kindergartners in FOSS, called **embedded assessment**, occurs on a daily basis. Teachers observe action in class or review notebooks after class. Embedded assessments provide continuous monitoring of students' learning and help you make decisions about whether to review, extend, or move on to the next idea to be covered.

The FOSS assessment system has other components for older students that are medium- and long-cycle. **Benchmark assessments** are medium-cycle summative assessments given after each investigation. These **I-Checks** are actually hybrid tools that provide summative information about student achievement, but because they occur soon after teaching each investigation, they can be used diagnostically as well. Reviewing a specific item on an I-Check with the class provides another opportunity for students to clarify their thinking. **Interim assessments** are designed to be more like district and state science tests, and focus on specific performance expectations from the NGSS, but can also be used as formative assessments for learning.

For the kindergarten modules, assessment is exclusively embedded, formative assessment—you observe students' actions in class.

Contents

The FOSS Assessment System for Kindergarten **233**

Embedded Assessment **234**

NGSS Performance Expectations for Kindergarten **238**

Summary of Assessment Opportunities in Kindergarten **240**

Full Option Science System

ANIMALS TWO BY TWO — Assessment

EMBEDDED *Assessment*

Assessment and teaching go hand in hand. Assessing students on a regular basis gives you valuable information to guide instruction and to keep families and other interested members of the educational community informed about students' progress. Assessment should be an ongoing part of everyday life in the classroom.

FOSS embedded assessments allow you and your students to monitor learning on a daily basis as you progress through the module. You will find suggestions for what to assess in the Getting Ready section of each part of each investigation. For example, here is a typical Getting Ready step for Part 1 of the first investigation.

15. Plan assessment for Part 1
Print or make copies of these assessment checklists to use throughout this investigation. Most of these checklist copies can be used for other investigations as well.
- *Assessment Checklist* 1: Disciplinary Core Ideas
- *Assessment Checklist* 2: Science and Engineering Practices
- *Assessment Checklist* 3: Crosscutting Concepts

What to Look For

- *Students handle living things with respect as they make and record observations of the structures of fish. (Planning and carrying out investigations; structure and function.)*

- *Students begin to construct the core idea that animals (fish) have exernal parts that help them meet their needs. (LS1.A: Structure and function.)*

Three assessment checklists are provided in the Assessment Masters to record observations about students' progress at the science center or in students' notebooks. Print or photocopy the sheets, and attach them to a clipboard so you can keep them nearby. In most cases, you will assess one science and engineering practice, one disciplinary core idea, and one crosscutting concept during each part of an investigation.

Embedded Assessment

Assessment Checklists

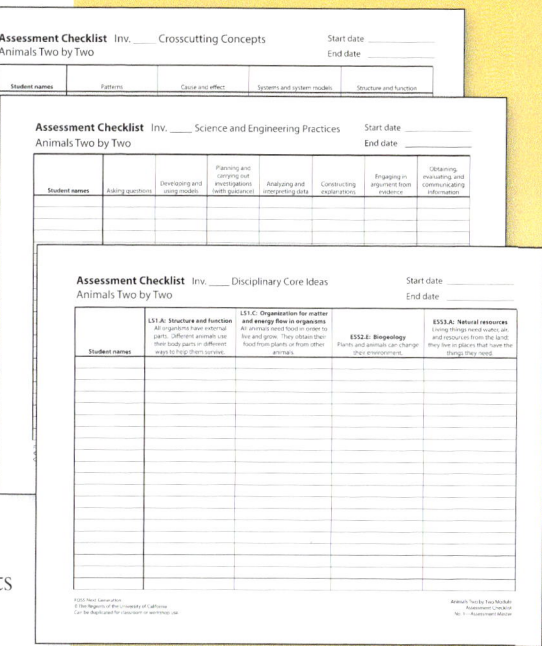

The first checklist (Assessment Master 1) is for recording students' progress on the disciplinary core ideas developed throughout the module. You can record on one sheet for all four investigations, or print or make copies of the appropriate sheets and use them for each investigation. There is space on each sheet to record the investigation number.

The second checklist is for recording progress in science and engineering practices. There are opportunities for observing students engaging in science and engineering practices throughout the module. The third checklist is for recording engagement in crosscutting concepts. See the following pages for more information about science and engineering practices as well as crosscutting concepts as described in the NGSS.

Observing at the science center. You will find that opportunities for students to develop the three-dimensional learning described above abound in the science center. The Getting Ready section for each part highlights those assessments in a particular activity. Focus on a few students at a time when using the checklist. Over the course of the module, you'll observe each student several times. Each time you observe a student, mark the assessment checklists with the date and a + or – to indicate progress. This system will also help you keep track of how often you have observed each student. Look for improvement over the course of the module.

Science notebooks. Making good observations and using them to develop explanations for how the natural world works are the essence of science. This process calls for critical thinking and good communication skills. Students' notebooks are a useful assessment tool and a good language extension. Use the notebook entries to provide you with evidence that students are accomplishing the learning goals of the module.

Nos. 1–3—Assessment Masters

Narrative Report

The Assessment Masters include an individual *Narrative Report* to send home to families or pass on to first-grade teachers. Print or make one copy of this two-page report for each student, and fill it out, using the records you have been keeping on the assessment checklists. In the comments section, write a few observations about the progress you have seen in each student's science learning from the beginning to the end of the module.

Nos. 4–5—Assessment Masters

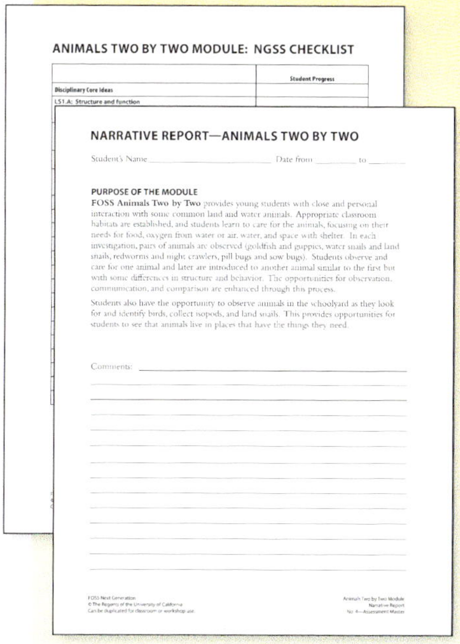

Animals Two by Two Module—FOSS Next Generation

ANIMALS TWO BY TWO — Assessment

Science and Engineering Practices for K from NGSS

1. **Asking questions and defining problems**
   - Ask questions based on observations to find more information about the natural and/or designed world(s).
   - Define a simple problem that can be solved through the development of a new or improved object or tool.

2. **Developing and using models**
   - Develop a simple model based on evidence to represent a proposed object or tool.

3. **Planning and carrying out investigations**
   - With guidance, plan and conduct an investigation in collaboration with peers.
   - Make observations (firsthand or from media) and/or measurements to collect data that can be used to make comparisons.

4. **Analyzing and interpreting data**
   - Record information (observations, thoughts, and ideas).
   - Use and share pictures, drawings, and/or writings of observations.
   - Use observations (firsthand or from media) to describe patterns in the natural world in order to answer scientific questions.

5. **Using mathematics and computational thinking**
   - Use counting and numbers to identify and describe patterns in the natural and designed world(s).
   - Describe, measure, and/or compare quantitative attributes of different objects and display the data using simple graphs.

6. **Constructing explanations and designing solutions**
   - Make observations (firsthand or from media) to construct an evidence-based account for natural phenomena.
   - Use tools and/or materials to design and/or build a device that solves a specific problem or a solution to a specific problem.

Embedded Assessment

7. Engaging in argument from evidence
- Construct an argument with evidence to support a claim.

8. Obtaining, evaluating, and communicating information
- Read grade-appropriate texts and/or use media to obtain scientific information to determine patterns in the natural world.
- Communicate information or solutions with others in oral and/or written forms using models or drawings that provide detail about scientific ideas.

Crosscutting Concepts for K from NGSS

Patterns
- Patterns in the natural and human-designed world can be observed, used to describe phenomena, and used as evidence.

Cause and effect
- Events have causes that generate observable patterns.

Scale, proportion, and quantity
- Relative scales allow objects and events to be compared and described (e.g., bigger and smaller; hotter and colder; faster and slower).

Systems and system models
- Objects and organisms can be described in terms of their parts.
- Systems in the natural and designed world have parts that work together.

Energy and matter
- Objects may break into smaller pieces, be put together into larger pieces, or change shapes.

Structure and function
- The shape and stability of structures of natural and designed objects are related to their function(s).

ANIMALS TWO BY TWO – *Assessment*

NGSS *Performance Expectations for Kindergarten*

"The NGSS are standards, or goals, that reflect what a student should know and be able to do; they do not dictate the manner or methods by which the standards are taught. . . . Curriculum and assessment must be developed in a way that builds students' knowledge and ability toward the PEs [performance expectations]" (*Next Generation Science Standards*, 2013, page xiv). The FOSS assessment system for kindergarten focuses on formative embedded assessments. The chart displayed on the next page provides an overview of these assessments across the three kindergarten modules. These assessments help students build knowledge and ability in concert with the active investigations and readings to meet the goals of the NGSS.

NGSS Performance Expectations for Kindergarten

Kindergarten NGSS Performance Expectations	FOSS Module Embedded Assessment
K-PS2-1. Plan and conduct an investigation to compare the effects of different strengths or different directions of pushes and pulls on the motion of an object.	Materials and Motion
K-PS2-2. Analyze data to determine if a design solution works as intended to change the speed or direction of an object with a push or a pull.	Materials and Motion
K-PS3-1. Make observations to determine the effect of sunlight on Earth's surface.	Trees and Weather Materials and Motion
K-PS3-2. Use tools and materials to design and build a structure that will reduce the warming effect of sunlight on an area.	Materials and Motion
K-LS1-1. Use observations to describe patterns of what plants and animals (including humans) need to survive.	Animals Two by Two Trees and Weather
K-ESS2-1. Use and share observations of local weather conditions to describe patterns over time.	Trees and Weather
K-ESS2-2. Construct an argument supported by evidence for how plants and animals (including humans) can change the environment to meet their needs.	Animals Two by Two Trees and Weather
K-ESS3-1. Use a model to represent the relationship between the needs of different plants or animals (including humans) and the places they live.	Animals Two by Two Trees and Weather
K-ESS3-2. Ask questions to obtain information about the purpose of weather forecasting to prepare for, and respond to, severe weather.	Trees and Weather
K-ESS3-3. Communicate solutions that will reduce the impact of humans on the land, water, air, and/or other living things in the local environment.	Materials and Motion
K–2-ETS1-1. Ask questions, make observations, and gather information about a situation people want to change to define a simple problem that can be solved through the development of a new or improved object or tool.	Materials and Motion
K–2-ETS1-2. Develop a simple sketch, drawing, or physical model to illustrate how the shape of an object helps it function as needed to solve a given problem.	Materials and Motion Trees and Weather
K–2-ETS1-3. Analyze data from tests of two objects designed to solve the same problem to compare the strengths and weaknesses of how each performs.	Materials and Motion

ANIMALS TWO BY TWO – Assessment

SUMMARY of Assessment Opportunities in Kindergarten

	Inv. 1	Inv. 2	Inv. 3	Inv. 4
Disciplinary Core Ideas				
LS1.A: Structure and function All organisms have external parts. Different animals use their body parts in different ways to help them survive.	✓	✓	✓	✓
LS1.C: Organization for matter and energy flow in organisms All animals need food in order to live and grow. They obtain their food from plants or from other animals.	✓		✓	✓
ESS2.E: Biogeology Plants and animals can change their environment.	✓	✓	✓	
ESS3.A: Natural resources Living things need water, air, and resources from the land; they live in places that have the things they need.	✓		✓	✓
Science and Engineering Practices				
• Asking questions	✓	✓	✓	✓
• Developing and using models	✓		✓	
• Planning and carrying out investigations	✓	✓	✓	✓
• Analyzing and interpreting data	✓	✓	✓	✓
• Constructing explanations	✓	✓	✓	✓
• Engaging in argument from evidence		✓	✓	
• Obtaining, evaluating, and communicating information	✓	✓	✓	✓
Crosscutting Concepts				
• Patterns	✓	✓	✓	✓
• Cause and effect	✓	✓	✓	✓
• Systems and system models	✓	✓	✓	✓
• Structure and function	✓	✓	✓	✓